Understanding Geography 3

People at Work

by Norman Pantling B.Sc.

Illustrated by
Geoff Bucktrout

Schofield & Sims Ltd Huddersfield

Preface

0 7217 1054 9

First printed 1983

Reprinted 1984, 1985

Acknowledgements
The author and the publishers wish to thank the following for permission to use copyright material and photographs:
J Allan Cash Ltd *4(3), 37, 45, 69(3)*
Marks and Spencer plc *4*
National Coal Board, Regional Services, Doncaster *4*
Austin Rover Group Ltd *4*
Derek Widdicombe, Countrywide Photographic Library *4, 38*
British Railways Board *4, 74*
International Wool Secretariat *4*
Popperfoto *4, 9, 17*
Aerofilms *11, 21, 23*
Airviews (Manchester) Ltd *14, 34, 79, 91*
Allied Bakeries Ltd *15*
British Sugar plc *18*
British Steel Corporation *20, 21*
Hypermarket (Holdings) Ltd *24*
United Kingdom Atomic Energy Authority *38*
British Petroleum *47, 95*
Central Electricity Generating Board, London *50*
Bureau of Reclamation, United States Department of the Interior *53, 62*
Topham *59, 60, 70*
Martyn Cowley, Gossamer Ventures *59*
High Commission for New Zealand *61*
Sólarfilma *61*
National Coal Board, Huthwaite, Nottinghamshire *69*
The Felixstowe Dock and Railway Company *74, 94(2)*
Peter Lockwood *78*
River and Gulf Transportation Company *89*
British Waterways Board *91(2)*
British Transport Docks Board, Swansea and Port Talbot *93*
British Airports Authority *97*

Understanding Geography is a series of five books designed to serve the needs of pupils in Comprehensive Schools. It offers a structured programme of learning with a variety of photographs, diagrams and data to illustrate the text. The approach is intended to capture and sustain the interest of pupils with differing motivation and ability, and may be used with mixed-ability classes or selective groupings.

Industry, Energy and Transport provide the major themes of this book, the third in the series. Chapter 1 examines the world of work and emphasises the changing patterns in the primary, manufacturing and service sectors. The causes and repercussions of unemployment (matters often neglected by geographers) are also considered in both economic and regional terms. In Chapter 2, renewable power resources and fossil fuels are discussed against a background of growing energy demands and threatened power shortages. Careful consideration is given to the benefits and dangers of nuclear energy, and alternative sources of power are described. Chapter 3 traces how new methods of transport are changing the pattern of personal journeys, goods traffic and public passenger travel. Current developments in rail, road, water and air transport are considered in relation to each other and in terms of their environmental impact. A topic of growing significance, transport networks, introduces several absorbing case-studies.

A series of carefully-worded exercises allows pupils to test their knowledge and understanding, and enables them to apply key geographical ideas to practical problems. Questions require specific answers which may be obtained from the text and its supporting illustrations.

This book is designed as a balanced scheme of work for pupils about to complete their geographical studies at the age of 14, whilst providing a valuable resource for those who opt for further studies leading to public examinations at 16+.

Norman Pantling

Printed by Eyre & Spottiswoode Ltd, Thanet Press, Margate

Contents

A

B

C

D

E

F

G

H

I

J

Fig. 1 People at work

Chapter 1 People at Work

Industry

To produce the food, shelter, clothing and other things we need, people work in a wide range of activities known as *industry*. All the workers seen in Fig. 1 are involved in some kind of industry. Some of these workers provide *raw materials*. Others use the raw materials to make the goods we need. Many workers transport raw materials to the factories and goods to the shops. Workers in shops and supermarkets sell the goods to the public. Look at Fig. 1 and try to identify the job of each worker.

Types of job

Industry may be divided into three groups or sectors:

1. Workers in *primary* industry obtain raw materials from nature. Examples of primary workers are farmers, forestry workers, miners and fishermen.

2. Workers in *secondary* industry make goods from raw materials supplied by the primary industries. For example, wheat is processed by *milling* to make flour. Bricks are *fired* from clay and petrol is *refined* from oil.

Some secondary industries use the products of other secondary industries. For example, a car factory *assembles* components – tyres, radiators, engines, batteries and all the other parts that go together to make a car. These components are supplied by other manufacturers.

Secondary industries may be broadly divided into two types:
(a) *Heavy* industries manufacture goods such as iron and steel, build ships and construct bridges.
(b) *Light* industries produce goods such as toys, watches, transistor radios, food mixers and many other similar goods.

Exercise 1 A nation's work-force

1 Study the following list of workers: bus driver, fisherman, dentist, baker, miner, insurance clerk, shop assistant, disc jockey, professional footballer, carpenter, market gardener, quarryman.
   Which are (a) primary workers; (b) secondary (manufacturing) workers; (c) service workers?

2 Rearrange each list of words to make sense.
   (a) raw industry primary materials supplies
   (b) products make manufacturers goods from other
   (c) employed British most workers are service in industries.

3 Study Fig. 1. Which of the workers pictured in A to J are service workers?

4 Place the following in order of processing. For example: grapes – wine – bottles.
   (a) bakery   wheat   farm   flour   mill   shop   bread
   (b) forest   house   builder   tree   sawmill   planks
   (c) house   clay   bricks   furnace

   (d) ship   refinery   pipeline   oil well   road-tanker   petrol station   car.

5 Which of the following statements does *not* help to explain why developed nations have a high proportion of service workers?
   (a) Rich nations can afford education, medical treatment and entertainment.
   (b) Rich countries use machinery rather than human effort.
   (c) Less developed countries are densely populated.

6 Study Fig. 2. Are the following statements true or false?
   (a) Most of the Indian work-force are primary workers.
   (b) Most workers in the USA are in service occupations.
   (c) Half of Britain's work-force is employed in manufacturing.
   (d) One in ten of Egypt's workers are employed in manufacturing industries.

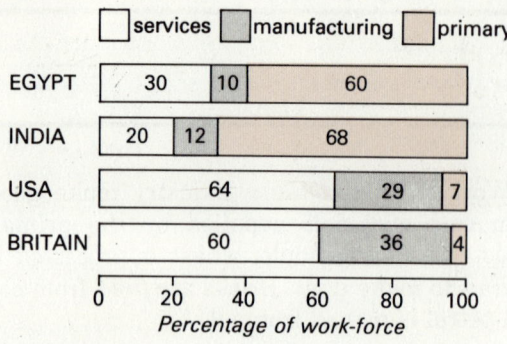

Fig. 2 The percentage of the work-force in each sector of industry for selected countries

3. The third group is the *service* industries. Workers in service industries do not make finished goods but provide essential support for other industries and for the public. These services include: transport; retailing (shops); banking; insurance; civil service and education. Fig. 2 shows that 60% of Britain's workers are employed in services. Less developed countries such as India have a much smaller percentage of service workers. Most Indian workers are farmers who use traditional methods of farming which require manual labour. As countries increase their living standards, primary and secondary industries become more mechanised and people require more services such as transport, education and entertainment.

Now try Exercise 1.

Primary industry: fishing

With a yearly world catch equal to two-thirds of the world's annual meat production, fishing is a good example of a major primary industry. Fig. 3 tables the output of the world's major producers. Although modern methods include radar-equipped factory ships, fishing is still a primitive way of obtaining protein-rich food. It is a form of hunting.

Fig. 4 shows the importance of fish in the Japanese diet. Farmers here are handicapped by steep mountains and a short growing season so people have turned to the rich fishing grounds which lie offshore. In the United States and Argentina, however, where conditions favour the production of low-priced meat, fish provides only a small part of the diet. Iceland, a remote, sub-arctic island in the North Atlantic (see Fig. 5), depends almost entirely on its fishing industry. Its barren icy wastelands contrast sharply with its offshore waters which are rich in fish. Fish provide four-fifths of Iceland's exports. It is no surprise, then, that in 1975 the Icelanders imposed a ban on fishing by foreign nations within 320 kilometres of the shore. Since then most major coastal fishing nations, including Britain, have imposed similar limits in order to protect fish stocks.

Fishing methods

Fish feed on tiny plants and animals known as *plankton*. Plankton grows best in the shallow waters of *continental shelves* (Fig. 6). Here sunlight can penetrate to the sea-bed and waters are richly supplied with oxygen and nutrients brought down by rivers. Fig. 5 shows the world's major fishing grounds. These are found where waters are rich in plankton. Note their wide extent in the Northern Hemisphere between 30° and 60° North.

Fishing methods depend upon the type of fish being caught. There are three main groups:
1. *Pelagic* fish such as herring, mackerel, sprat and pilchard swim near the surface, often in huge shoals. They are usually caught by *drift-* or *seine-nets* (Fig. 7). Drift-nets are weighted and hang down through the water from a line of surface floats. Fish are caught by their gills in the narrow mesh of the net. Seine-nets, similar to drift-nets in size and mesh, are first towed around a shoal before ropes along the bottom of the net are drawn in to make a 'purse' in which the fish are caught.
2. Fish which live on or near the sea-bed, such as cod and plaice, are known as *demersal* fish and are usually caught by a *trawl-net* (Fig. 8). This is shaped like a cone and is dragged along the sea-bed by a vessel called a *trawler*. In Japan and Newfoundland, cod are still caught by baited lines, often two kilometres long, which are trailed from small boats.
3. *Shellfish* such as lobster, crab, crayfish and Norwegian lobster or 'scampi', are taken from shallow inshore coastal waters. They are caught in baited baskets or *pots* and then collected by small boats. This costly method of fishing is reflected in the high shop and restaurant prices obtained for such delicacies.

Now try Exercise 2.

World fish catches (1978)

	thousand tonnes
Japan	10 752
USSR	8930
China	4660
USA	3512
Peru	3365
●Norway	2647
India	2368
S. Korea	2351
Thailand	2264
●Denmark	1745
Chile	1698
Indonesia	1655
N. Korea	1600
Iceland	1579
Philippines	1588
Canada	1407
●Spain	1380
●Britain	1054
Vietnam	1014
World total	**72 380**

●European nations

Fig. 3 The major fishing nations

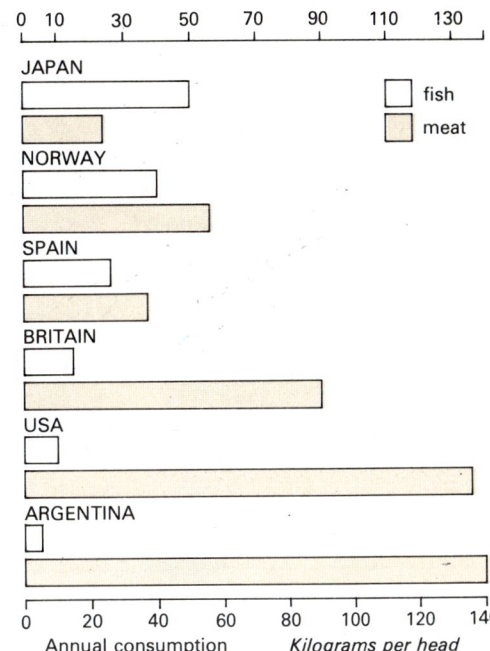

Fig. 4 The fish and meat consumption of selected countries

Fig. 5 The world's major fishing grounds

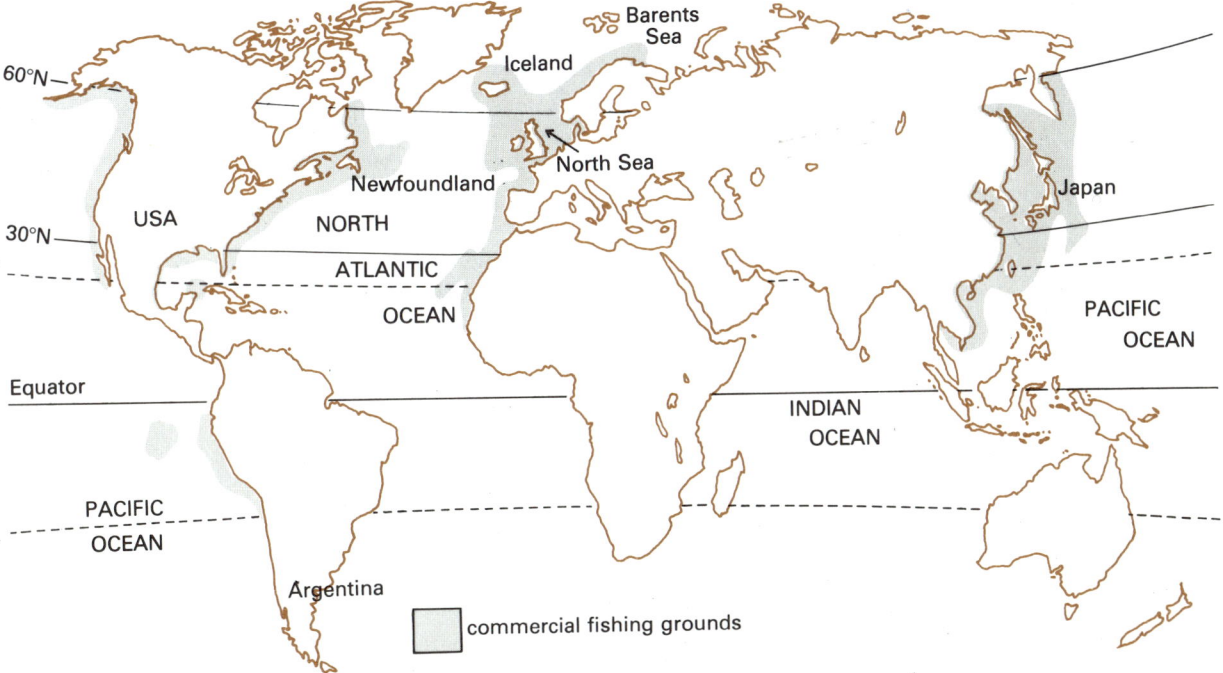

commercial fishing grounds

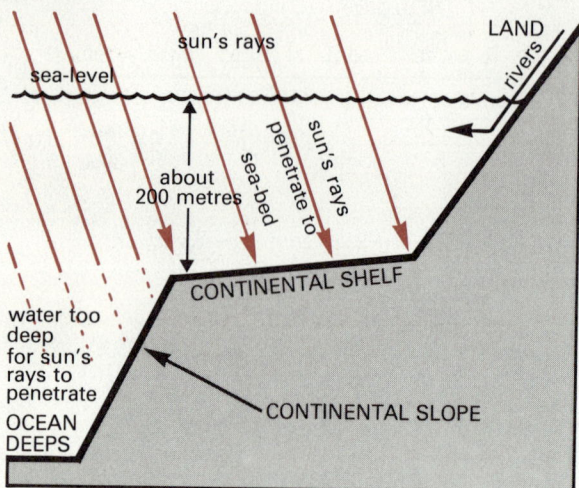

Fig. 6 A cross-section of the continental shelf

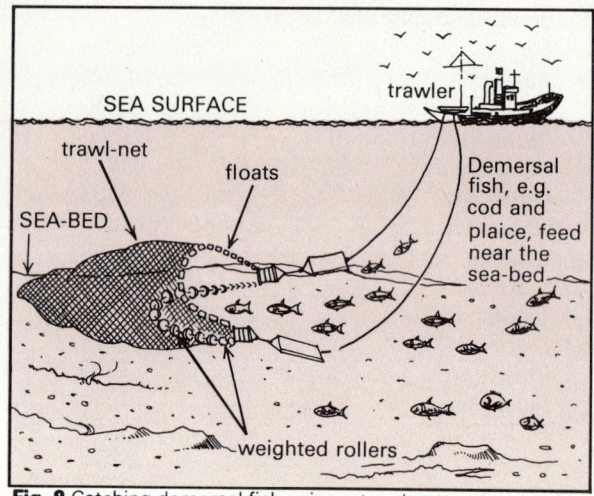

Fig. 8 Catching demersal fish using a trawl-net

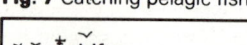

Fig. 7 Catching pelagic fish

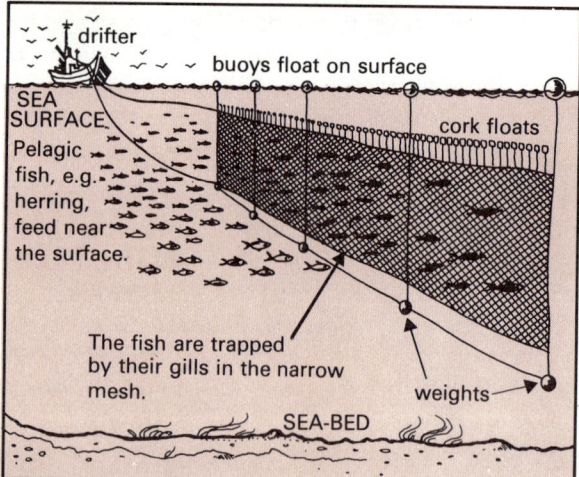

a Using a drift-net

b Using a seine-net

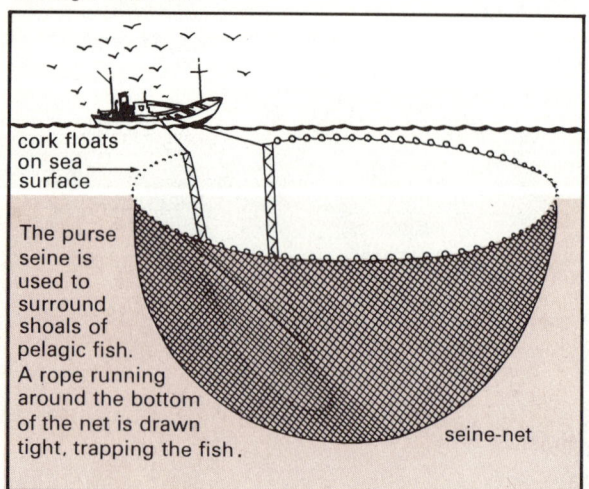

Exercise 2 The fishing industry

1 Give two reasons for the importance of fishing in Japan.
2 What step have Icelanders taken to protect their fishing industry?
3 What is plankton?
4 Describe two factors which favour the growth of plankton.
5 What term is used to describe shallow offshore waters?
6 Why is a trawl-net unsuitable for catching herring?
7 To which group of fish does cod belong?
8 Describe two ways of catching cod.
9 Name two islands in the North Atlantic Ocean where fishing is important (excluding the British Isles).
10 Name the group (pelagic, demersal or shellfish) to which each of the following fish belongs: plaice, lobster, crab, mackerel, pilchard, herring, halibut.

Overfishing

Between 1950 and 1978 the world fish catch increased steadily from 21 million tonnes to 72 million tonnes (Fig. 9). Increasing competition for what appears to be the ocean's 'free gift' has in recent years resulted in the near exhaustion of certain species of fish. Such has been the fate of the anchovy off the coast of Peru, and the Californian sardine. In the past, catches were limited by the small size of vessels and the comparatively short voyages made. Fish stocks were therefore replenished naturally and could be harvested year by year as a renewable resource. Today, however, fish stocks are threatened by the growing number of highly efficient vessels and the lack of effective international control on the size of their catch.

The modern diesel-engined trawler is able to withstand the buffeting of violent seas and the dangerous icing-up which occurs in arctic waters. Huge trawl-nets are hauled mechanically through the stern into the safety of enclosed decks where the catch is gutted, cleaned and stored in deep-freeze holds. One hundred and twenty tonnes of fish can be processed each day and the voyages often last several weeks. Vessels equipped with radio, radar and electronic acoustic equipment can detect and trap a 600-tonne shoal of mackerel in a single seine-purse net 1·5 kilometres long. Even larger factory ships (Fig. 10) supplied by a fleet of smaller craft can process more than 3000 tonnes of small pelagic fish each day into meal used to feed chickens and pigs.

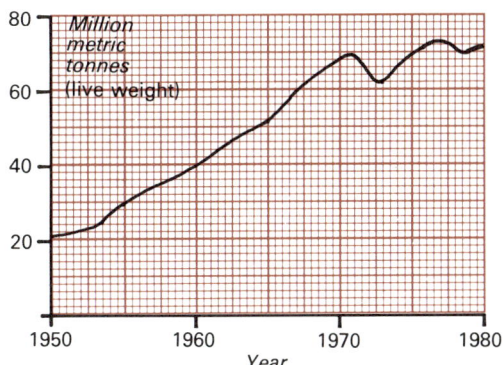

Fig. 9 The world fish catch from 1950 to 1978

Uncontrolled fishing on a large scale threatens the future of the industry. Overfishing occurs when the total catch of a particular species leaves too few mature fish behind to breed and maintain stocks. This danger is increased by the use of nets with a mesh too fine to allow even infant fish to escape. Furthermore, scientists who monitor the fish population may not discover real proof of overfishing until the damage has been done and it is too late to remedy.

Fig. 11 gives recent variations in the herring catch in Britain. Total landings fell steadily from more than 260 000 tonnes in 1948 to 91 000 tonnes in 1976 when scientists warned that stocks were seriously depleted. Since the introduction of severe controls on herring fishing in 1977, stocks have recovered sufficiently to allow a catch of over 36 000 tonnes in 1981.

Fig 10 The Russian fish factory ship *Biosfera*

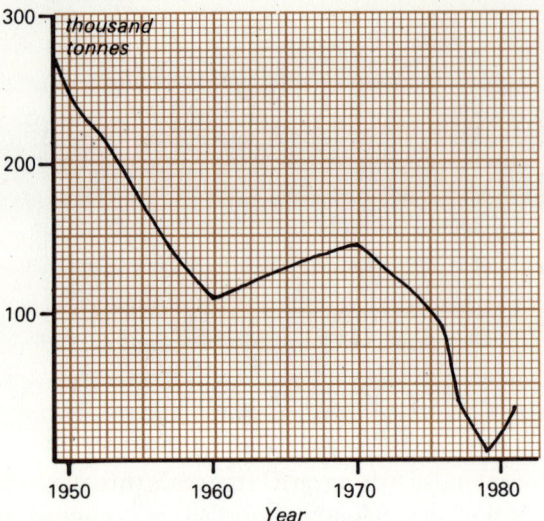

Fig. 11 The total herring catch landed by British vessels from 1949 to 1981

In 1977, following the United Nations Conference on the Law of the Sea, international agreement was reached which enabled coastal nations to establish their own Exclusive Economic Zones (EEZ). This gives each nation sovereign rights to explore and exploit the natural resources of the sea and sea-bed up to 320 kilometres from the shore. Where neighbouring states are separated by waters less than 640 kilometres wide, as for example in the North Sea, frontiers are normally agreed along a median (middle) line. Although the creation of Exclusive Economic Zones helps nations to protect fish stocks within their own territorial waters, several traditional fishing grounds, such as Icelandic waters, have been closed to British and other foreign vessels, forcing them to concentrate even more upon their own coastal waters. As a result, increased pressure has been placed upon North Sea fish stocks.

Without international co-operation and controls, world fish reserves could quickly dwindle to dangerously low levels. A declining fish catch would then increase pressure on land-based food supplies. Although careful management of traditional species such as cod and herring is now widely recognised, effective international regulations have yet to be introduced. The exploitation of non-traditional species such as blue whiting and scad is now under investigation. Krill, a small shrimp-like crustacean which was once the plentiful food of antarctic whales, is found in gigantic quantities in the Southern Ocean. Japanese, Russian and West German factory ships have recorded large catches since 1976. The exploitation of such unfamiliar species is not merely a challenge for fishermen, but also for the retailer, who must persuade customers to accept the unfamiliar flavours before major investment takes place.

Now try Exercise 3.

Exercise 3 Overfishing

Answer true or false.

1 The world fish catch trebled in size between 1950 and 1978.
2 Stocks of herring and anchovy have been threatened by overfishing.
3 Nets of fine mesh help to preserve fish species.
4 Deep underwater shoals of fish can be detected by acoustic equipment.
5 Setting up an Exclusive Economic Zone (EEZ) permits foreign vessels to fish territorial waters.
6 The EEZ have increased pressure on Icelandic fish stocks.
7 The EEZ have reduced pressure on North Sea fishing grounds.
8 Krill is an abundant pelagic fish in the Southern Ocean.
9 World fish stocks are now carefully managed.
10 In 1978 Japan and the USSR landed more than 25% of the world's fish catch.

Fishing in Britain

Once the world's greatest fishing nation, Fig. 3, page 7, reminds us that Britain remains one of Europe's most important. The industry provides 70% of the nation's fish consumption and gives employment to nearly 17 000 full-time fishermen and to 40 000 workers in services such as fish processing and distribution.

Fig. 12 shows the British Isles surrounded by an extensive continental shelf. Its shallow sunlit waters favour the growth of plankton which support a wide variety of fish species. In 1980, 90% of the total British catch was taken from these home waters. The major fishing ports are mapped in Fig. 13.

Demersal fish accounted for 54% of the total weight caught by British vessels in 1980. Cod is by far the most popular fish, accounting for 31% of total sales in 1980, with haddock 18%, mackerel 13% and plaice 7%, all important sources of revenue.

Since 1960 there have been notable changes in the way fish is prepared for the consumer. One third of

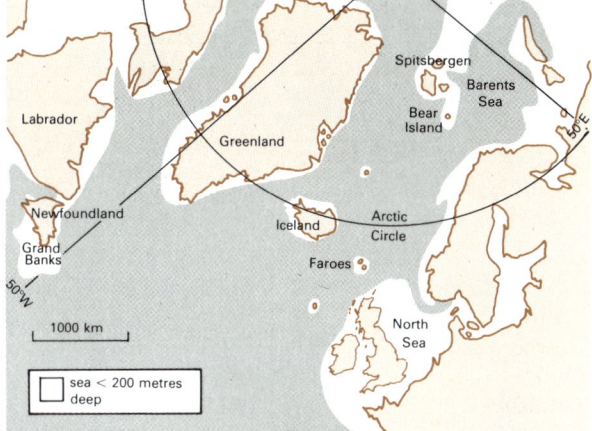

Fig. 12 The fishing areas of the North Atlantic

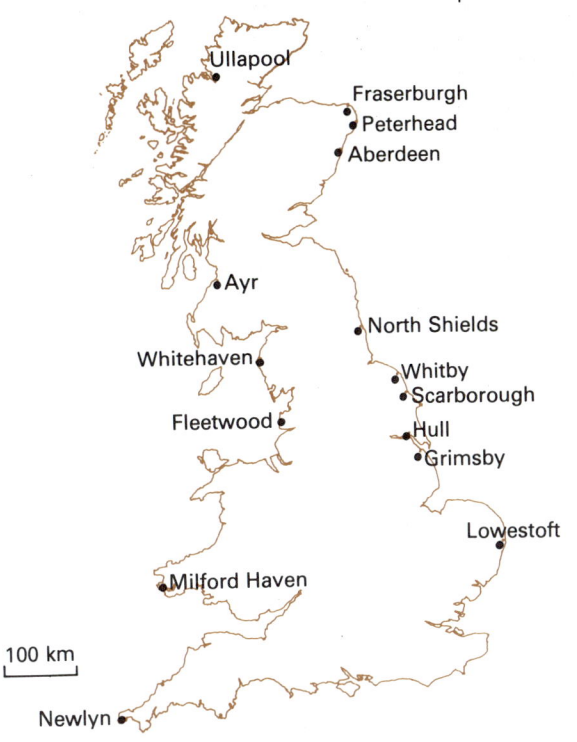

Fig. 13 Britain's major fishing ports

Fig. 14 The large Bird's Eye factory at Grimsby where fish is processed and packaged for sale at supermarket frozen food counters

all fish sold is now bought as a frozen product, with declining amounts purchased from fishmongers and fish and chip shops. Since 1972, 50% of fishmongers have ceased trading. International companies such as Ross, Findus and Bird's Eye, whose factory at Grimsby is seen in Fig. 14, have transformed the way we buy fish. In the past, all fish was auctioned to dealers at quayside or at specialist markets such as Billingsgate in London. Packed in ice, it was quickly despatched to fishmongers for sale as a fresh product; other fish were preserved (and some still are) by smoking or pickling, or sold as a salted or dried fish. Today, shoppers increasingly prefer the stable price and convenience of the frozen product which is processed, packaged and delivered to supermarkets by the same firms that own the trawlers.

Fig. 15 reveals the steady decline of the modern fishing industry in Britain since 1948. Serious problems of overfishing, loss of traditional fishing grounds to the 320-kilometre limit, and a gradual fall in demand have created serious unemployment in specialist ports. Fig. 16 tables the changing fortunes of major centres. The catastrophic decline in landings recorded by Hull, Grimsby and Fleetwood since 1970 reflect the loss of deep-sea fishing grounds off Iceland, Greenland, Newfoundland and the Faroes. Ullapool has not fared so badly. Its increased catch of 108 000 tonnes in 1981 was comprised almost wholly of mackerel taken from newly-developed grounds off the west coasts of Scotland and Ireland.

Although the European Economic Community (EEC) was established in 1958 (Britain joined in 1973), a Common Fisheries Policy has yet to be decided. Long term aims, however, are firstly to conserve fish stocks by controlling mesh size, by excluding non-EEC vessels from home waters and by fixing a maximum Total Allowable Catch. How the catch is to be divided amongst the member nations, who are able to fish one another's waters, has still to be agreed upon. Britain has a particular interest because 60% of the Total Allowable Catch is taken from her extended territorial waters. The Common Fisheries Policy also aims to bring the total EEC catch into line with demand. Supplies will be controlled by a quota system designed to provide stable prices for customers whilst giving a reasonable return to producers.

	Total Landings *thousand tonnes*	Full-time Fishermen *thousands*	Annual Consumption *kg per head*
1948	1097.4	38 826	13.0
1960	843.4	22 007	7.3
1970	932.8	17 628	7.3
1980	759.1	16 716	5.4

Fig. 15 The decline of the British fishing industry

thousand tonnes	Demersal			Pelagic
	1960	1970	1981	1981
North Shields	13.3	39.8	13.8	0.6
Whitby	4.1	1.0	3.8	—
Scarborough	3.6	4.5	5.4	—
Hull	234.5	197.8	15.2	7.9
Grimsby	176.8	166.3	41.1	0.4
Lowestoft	22.8	28.6	18.4	—
Newlyn	3.9	3.1	12.8	7.4
Milford Haven	8.8	3.8	15.1	12.2
Fleetwood	50.4	41.2	11.9	—
Whitehaven	3.3	5.5	1.4	0.2
Aberdeen	90.7	120.9	42.0	—
Peterhead	10.9	12.0	82.0	—
Fraserburgh	24.5	18.6	34.7	17.1
Ullapool	22.8	46.0	108.0	94.6
Ayr	14.3	14.4	11.0	2.5

Fig. 16 Landings of fish at major fishing ports by British vessels.

The British fishing industry faces an uncertain future. The need for Government assistance, amounting to £40 million in 1981, is likely to continue as the industry adjusts to the loss of traditional fishing grounds and trims to take only the catch allowed by the EEC.

Now try Exercise 4.

(Another major primary industry, *coal-mining*, is described in Chapter 2).

Secondary industry

A secondary (or manufacturing) industry makes a product for sale. It may be a finished product or *consumer good* such as a loaf of bread, or it may be a product, such as steel bars or lengths of timber, that is further processed by other manufacturers. These are known as *capital goods*.

Manufacturing consists of the three main stages:
1. collecting raw materials;
2. processing raw materials;
3. distributing the product to customers.
These stages are illustrated by Fig. 17.

Making a profit?

Goods are made in the hope that customers will buy them at a price which will give the manufacturer a *profit*. Profit is what remains from the value of sales when all a factory's costs have been paid. Manufacturers try to keep costs as low as possible. Fig. 17 shows the costs which have to be paid by manufacturers. Raw materials must be bought and transported to the factory – these are known as *collection costs*. Making the finished product from these raw materials involves *processing costs* which include fuel bills and wages. Getting the completed article to the customer incurs *distribution costs* such as packaging, transport and advertising.

Some costs must be paid whether or not a factory is working at full capacity. These include interest paid to banks on borrowed money and rates due to the local authorities. Such payments are known as *fixed costs,* and must be made regardless of the profitability of the firm.

Fig. 17 The three main stages of a manufacturing industry

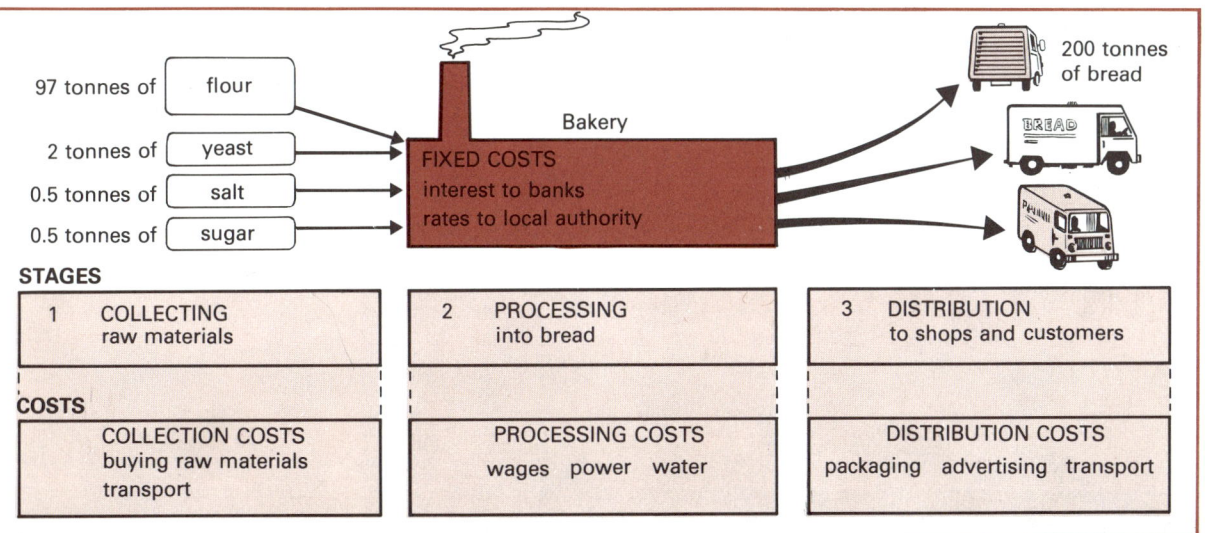

The location of manufacturing industry

Bread is made in thousands of bakeries spread across the nation. Steel is made in just a few parts of the United Kingdom. Factories which make sugar from beet are clustered in the eastern counties. What, then, determines the location of a particular manufacturing industry?

There are four main factors:

1. Industries such as baking, brewing beer, and newspaper printing benefit from a location close to their customers and are known as *market-based industries*.

2. Smelting aluminium requires vast quantities of cheap electricity, so plants are frequently situated near to the power source. This is an example of a *power-based industry*.

3. Fruit canning and the freezing of fresh vegetables is usually carried out in factories built near to crop-growing areas. Fresh produce is easily damaged. To preserve quality, the crop must be processed within hours of picking. Factories like these are located near to the source of the major *raw material*. Factories using bulky materials are also located near to the necessary raw materials to cut down heavy transport costs.

4. For some manufacturers, obtaining the right kind of *labour* is most important. An industrialist hoping to set up a cutlery-making factory would probably fail to obtain the right kind of labour in, say, Cheltenham or Eastbourne but the Sheffield district, with its long tradition of fine metal crafts, would provide a skilled work-force. Similarly, a manufacturer of sailing-boats would find the Southampton area a better location for a factory than, say, Blackpool or Aberystwyth. Within the Solent region live not only thousands of skilled craftsmen but also many potential customers. A boat factory here would reap the benefits of both availability of labour and closeness to a market for its product.

Now try Exercise 5.

1. Market-based industry

Some factories are best situated near to their customers. The cost of transporting a heavy, bulky, finished product may be much higher than the cost of collecting the raw materials together. The brewing of beer is a good example. Breweries are located in or near to towns to reduce the cost of transporting the heavy barrels to public houses and hotels (Fig. 18). Local evening newspapers are printed in the towns where they are circulated so they can be delivered with speed to street-sellers and newsagents. Local news can then be read within hours of its happening.

Fig. 18 The Ind Coope brewery at Romford supplies the company's many public houses in the London area

Exercise 5
Factory location and costs

Answer true or false.

1 Bars of steel are capital goods.
2 Bars of soap are consumer goods.
3 A van-driver's wages form part of a bakery's collection costs.
4 Fuel used by blast-furnaces is a distribution cost.
5 Buying raw materials is a fixed cost.
6 Market-based industries are often situated in or near large cities.
7 Freezing fresh peas is a power-based industry.
8 Breweries and bakeries are often built in rural areas close to raw materials.
9 Steel is made in most British towns.
10 Products which contain steel are made in most British towns.

Exercise 6
Market-based industry

Copy and complete the passage using the seven correct words chosen from the list:

customers, expensive, primary, towns, service, water, salt, gained, yeast, perishable, lost, collection, delivery, countries, market-based.

The baking of bread is a widespread _____ industry. Much weight is _____ in the baking process because _____ is added to the mix. Bread is a bulky and _____ product which must be sold to _____ while fresh. To keep _____ costs low, bakeries are found in most _____ .

Bread manufacture

The manufacture of bread is a good example of a market-based industry. The widespread location of factories of a leading manufacturer is mapped in Fig. 19. In addition to such large-scale manufacturers, bread is supplied by dozens of small bakeries in every city. It is a daily purchase which must be offered fresh in the shops. To reduce distribution costs and to speed delivery of this bulky, perishable commodity, bread is rarely transported long distances.

The bread-making process is shown in Fig. 17 (page 13). Note that the total weight of the raw materials – flour, sugar, salt and yeast – is half that of the bread made. Weight-gain is caused by the addition of water to the mix. Although water is a vital ingredient, it is widely and cheaply available and its supply does not influence the location of the bakery. Sixty percent of the wheat needed for baking bread is transported thousands of kilometres from North America. By contrast most loaves travel less than 50 kilometres from bakery to customer.

Now try Exercise 6.

Fig. 19 The market location of the factories of a leading bread manufacturer

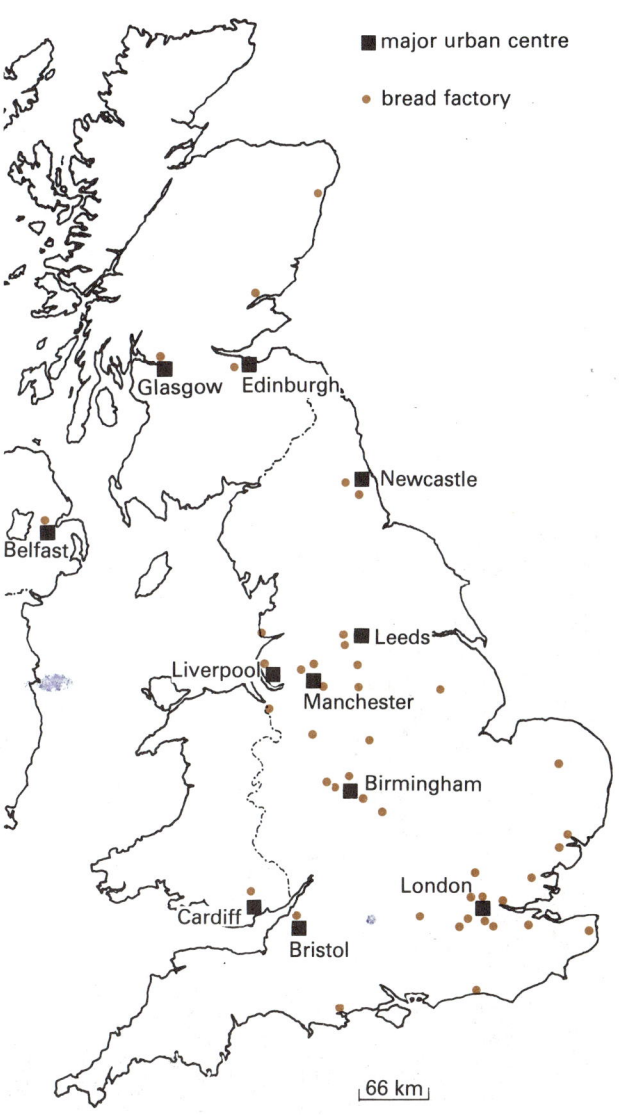

■ major urban centre

● bread factory

66 km

2. Power-based industry: Smelting aluminium

One of the largest but most isolated factories in North America is situated on the north-west coast of British Columbia (Fig. 20). The Kitimat refinery produces aluminium, but is situated away from its customers and far from its suppliers of raw materials. It is a power-based industry; that is to say, the most important factor of production is the availability of a cheap, plentiful power supply. The high rainfall and tall mountains of British Columbia favour the development of hydro-electric power (see page 53), and this is carried by *transmission lines* from Kemano power-station to Kitimat. Thus, the Kitimat aluminium plant has a reliable and cheap supply of power. It also has the advantage of being near the coast. The deep waters of the Douglas Channel provide a valuable harbour for vessels bringing raw materials and for exporting aluminium.

Aluminium is light in weight, does not corrode easily and conducts both heat and electricity. When manufactured, it is moulded into tubes and bars or rolled into sheets. Aluminium forms strong and durable alloys when mixed with other metals. It is widely used in aircraft and vehicle manufacture, and appears in such varied forms as tubular furniture, window frames, kitchen utensils and cooking-foil.

The raw materials and processes for making aluminium are shown in Fig. 21. This versatile metal is obtained by refining a clay-like material called *bauxite,* which is dug out of the ground from huge quarries (Fig. 22). Ten kilograms of raw bauxite produces one kilogram of aluminium. To extract the metal, the ore is first treated with powerful chemicals to remove some of the waste. The resultant white powder, called *alumina,* then undergoes a final process which requires enormous quantities of electricity. Alumina is dissolved in a chemical bath of molten cryolite. An extremely powerful electric current is then passed through this mixture to separate the molten aluminium which runs from the tanks into moulds. To keep costs low this process operates on a large scale twenty-four hours a day. The power needed to make just one kilogram of aluminium would keep a single bar electric fire going for 17 hours or a 100-watt light bulb for a week. It is

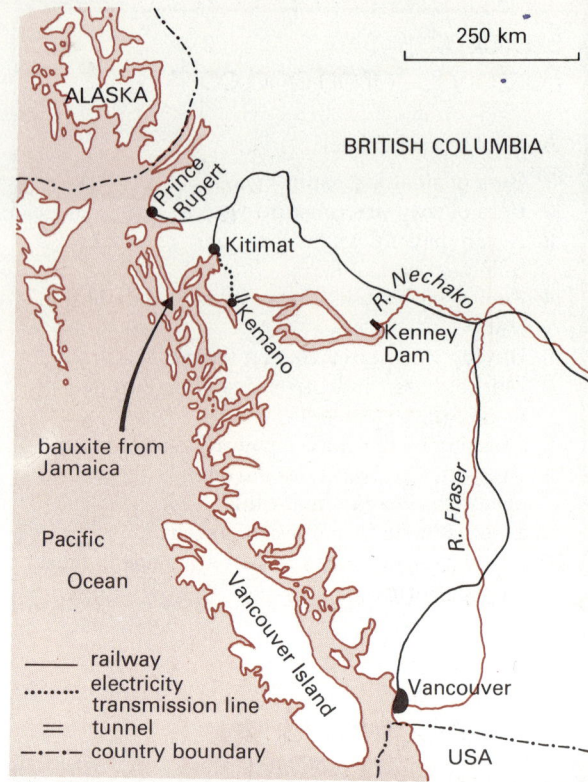

Fig. 20a The location of the Kitimat scheme

Fig. 20b A cross-section of the Kitimat scheme

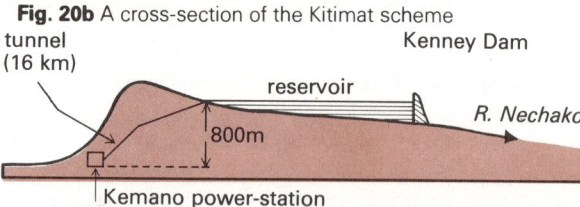

Fig. 20c A diagrammatic view of the Kitimat scheme

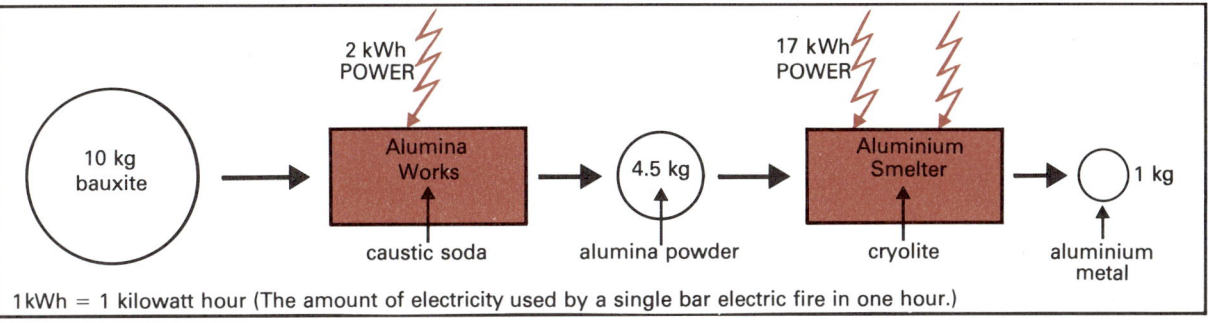

1kWh = 1 kilowatt hour (The amount of electricity used by a single bar electric fire in one hour.)

Fig. 21 The main stages of manufacturing aluminium

Fig. 22 Bauxite being quarried in Jamaica which produces 12 million tonnes per year

therefore easy to see the importance of locating an aluminium plant near to a good power supply.

Now try Exercise 7.

3. Raw material location: Refining sugar-beet

Britain uses 15 million tonnes of sugar each year. Sprinkled on cornflakes, and spooned into tea and coffee, sugar is also a major ingredient of cakes, biscuits, canned drinks and jams. Sugar has two main sources. Imported *cane sugar* is extracted from a tropical bamboo-like plant grown in the West Indies, South America, Africa and Australia. Cane sugar refineries are, therefore, located near the ports through which raw cane is imported.

Half of Britain's sugar supply, however, comes from home-grown *sugar-beet* – a turnip-like root which grows well in eastern England (Fig. 23).

Exercise 7 Power-based industry

1 Place the following words in order, starting with the raw material and ending with the manufactured product: cooking-foil, bauxite, ingots, alumina, rolling-mill.
2 How much bauxite is needed to make 100 tonnes of aluminium metal?
3 What source of power is used to make aluminium at Kitimat?
4 How much electricity is needed to make 10 kilograms of aluminium?
5 Give two reasons why Kitimat is a good location for aluminium production.

6 Are the following statements true or false?
 (a) Aluminium refining is a weight-gain process.
 (b) Kitimat is situated close to sources of bauxite.
 (c) Bauxite mining is a necessary manufacturing process.
 (d) The Kenney Dam holds back the waters of the Douglas Channel.
 (e) The Kemano power-station is situated beside the Kenney Dam.

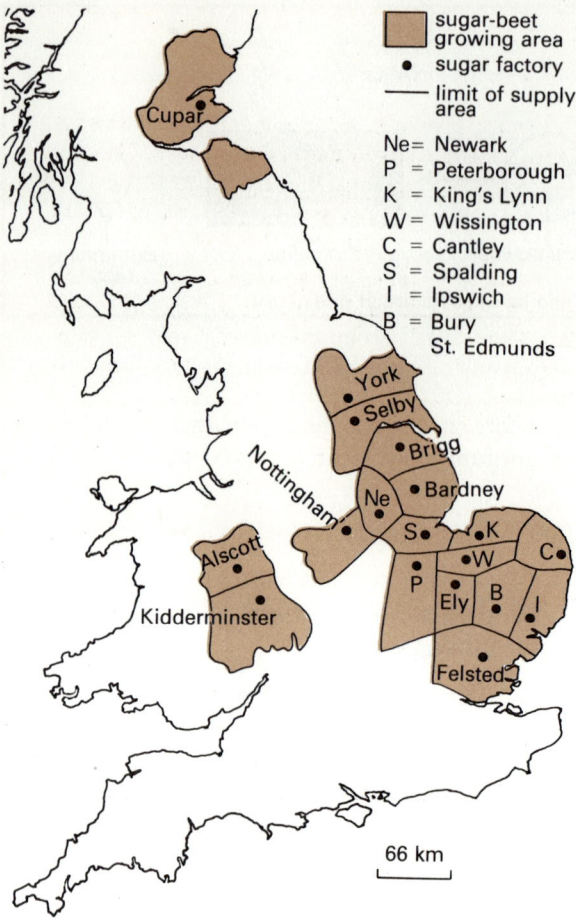

Ne = Newark
P = Peterborough
K = King's Lynn
W = Wissington
C = Cantley
S = Spalding
I = Ipswich
B = Bury St. Edmunds

66 km

Fig. 23 Sugar-beet growing areas and processing factories

Fig. 24 Piles of sugar-beet awaiting processing at a factory

Here, the gentle slopes, fertile soil and warm sunny summers give high yields per hectare.

The swollen roots seen in Fig. 24 contain juices from which sugar is extracted. They are transported by lorry to processing factories where they are sliced and crushed between heavy rollers. The juice is boiled to evaporate unwanted water. Fig. 25 shows how much weight is lost during the process. For every eight tonnes of beet only one tonne of sugar is produced. The rest of this weight is waste made up of water and pulp. Transporting

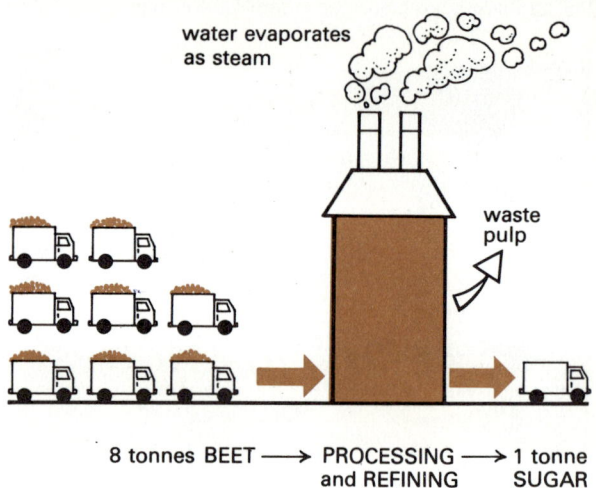

8 tonnes BEET ⟶ PROCESSING and REFINING ⟶ 1 tonne SUGAR

Fig. 25 For every eight tonnes of beet only one tonne of sugar is produced

Exercise 8 Choose a beet factory

A sugar manufacturer must process an 80-tonne load of beet in order to get 10 tonnes of sugar. The beet is grown at Green Farm (see map) and the refined sugar must be taken to the town of Stanton, 8 kilometres away.

The cost of transporting the total load of beet from Green Farm = £4 per km
The cost of processing the total load of beet = £50
The cost of transporting the 10 tonnes of refined sugar from the factory = 50p per km

There are three beet factories in the area. Factory A is right next to Green Farm, B is situated between the farm and Stanton and C is situated in the town itself. To keep costs as low as possible, which factory should be chosen to process the beet?

The following table can be used to calculate the costs and to discover the best choice. The total costs for factory B have been calculated. Copy and complete the table and then write down which factory is the most economical choice.

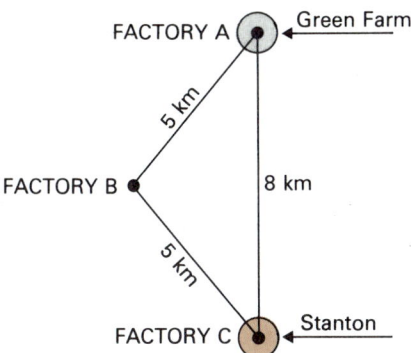

Factory	Costs			
	Transport of beet	Processing beet	Transport of refined sugar	Total Cost
A	NONE	£50		
B	£4×5km =£20	£50	50p×5km =£2.50	£72.50
C		£50		

Exercise 9 Sugar refining

1 From which two plants is sugar obtained?
2 What percentage of Britain's sugar is obtained from imported sugar cane?
3 How much beet is needed to make five tonnes of sugar? (answer in tonnes)
4 How much sugar is obtained from 24 tonnes of beet? (answer in tonnes)
5 Place the following in order of processing: packing, crushing, harvesting, slicing, boiling, delivery.

6 Name two waste products released by beet factories.
7 Choose two reasons from the following that help to explain why beet factories are built in growing areas:
   (a) Beet contains a high percentage of water.
   (b) Refined sugar is expensive to transport.
   (c) Beet bruises easily in transit.
   (d) Eight tonnes of beet make only one tonne of sugar.

raw beet is therefore very costly – most of the load ends up as waste. In order to cut costs, journeys to the factory are kept as short as possible. Beet factories are therefore built close to farms. Their location within the growing areas is mapped in

Fig. 23. The extracted raw sugar is sent by road to wholesalers in the major towns. The industry is an example of a *raw material location*. Exercise 8 demonstrates the advantage of such a location.

Now try Exercises 8 and 9.

←————IRON SMELTING————→ | ←————CONVERSION TO STEEL————→ | ←————FINISHING————→

coke

iron-ore

limestone

Blast-furnace

liquid pig-iron

Open-hearth Process

or

Basic Oxygen Converter

or

Electric Arc Furnace

additions:

steel scrap and other metals e.g. chromium, tungsten

range of steels

Rolling-mill

strip
plate
sheet
bars
beams
tubes

to customer

☐ process
☐ raw material
☐ product

Fig. 26 The main stages of steel making

Raw material location: Manufacturing steel

Steel manufacture is a *basic industry* upon which many other industries depend. Without steel there would probably be no large ships, tall buildings, railways, cars, aircraft or engines. Even the 'tin' can is made from a thin sheet of steel coated with tin. Factories producing millions of tonnes of steel each year are situated in places where a profit is likely to be made. A clear advantage is gained by building steelworks close to sources of raw materials since the heavy, bulky coal, iron-ore and limestone are more costly to transport than the finished steel.

Fig. 26 outlines the steel-making process. Raw materials, as with any other form of manufacture, must first be gathered together. Coke is made from coal in separate ovens to provide a suitable fuel. Coke is strong enough to support the immense weight of overlying materials yet porous enough to allow the passage of hot gases through the furnace. Thousands of tonnes of iron-ore, coke and limestone are tipped each day into huge blast-furnaces, such as the one in Fig. 27. At temperatures over 1800°C molten iron trickles to the bottom of the furnace. Every so often the metal is allowed to run out to be collected for the next stage of production. Some of the iron (now known as *pig-iron*) is set aside to make *cast-iron* grids, baths and engine parts. Most of the iron, however, is put into a *steel-furnace*. The three main types of steel-furnace are:

1. *The open hearth furnace*–for many years the most commonly used;
2. *The basic oxygen converter* – a quicker, cheaper method;
3. *The electric arc furnace* which is ideal for making small quantities of high-grade steel.

Customers such as car manufacturers, bridge-builders and aircraft engineers need steels of different strength, hardness and resistance to rust. To make different kinds of steel, furnace operators carefully adjust the carbon content of the molten metal, and add small quantities of other metals such as chrome, nickel or tungsten. Samples are taken until the furnace mix is ready for pouring into ingot moulds. Finally the hot soft ingots are converted by *rolling-mills* into a variety of bars, rails, sheets, etc. (Fig. 28).

Fig. 27 A blast-furnace used for making iron

Integrated steelworks

Blast-furnaces, steel-furnaces and rolling-mills may operate alone in separate sites, but the most modern *integrated works* combine all stages of production to save transport and fuel costs (Fig. 29). Operations are extremely *large scale*. Each day a typical blast-furnace uses 6000 tonnes of iron-ore, 2200 tonnes of coke and 400 tonnes of limestone to make 3500 tonnes of iron – which eventually provides 2800 tonnes of finished steel. The ratio of raw materials and fuel to finished products is just over 3:1, which means that steel-making is a *weight-loss* process. This is why steelworks were originally built close to supplies of the raw materials necessary for steel production.

Japanese steel

Where raw materials can be carried in modern ocean-going bulk-carriers, transport costs are kept low. Today, the world's largest steelworks are situated in coastal areas which have easy access to imported raw materials. They provide a useful example of *port-based* industry. The giant Nagoya

Fig. 28 A rolling-mill where steel is rolled into a variety of shapes

Fig. 29 The integrated steelworks at Port Talbot, S. Wales

Fig. 30 The location of the Nagoya steelworks

Exercise 10 Making steel

1 Why is steel manufacturing known as a *basic* industry?
2 Which raw material is processed in a coke oven?
3 Why must coke be (a) strong; (b) porous?
4 Place the following in order of processing: iron-ore, rolling-mill, oxygen converter, blast-furnace, pig-iron, steel girders.
5 Why is tungsten or chrome sometimes added to steel?
6 In which kind of furnace would small quantities of high grade steel be made?
7 In which part of the iron and steel manufacturing process is most weight lost?
8 What is an integrated steelworks?
9 Give two advantages of an integrated works.
10 Which one of the following statements does *not* help to explain the location of the Nagoya steelworks in Japan?
   (a) Iron-ore is imported.
   (b) Japan exports much steel.
   (c) The Nagoya works needs an extensive area of flat land.
   (d) Imported raw materials weigh much more than finished steel goods.

works in Japan (Fig. 30) is situated on the southern shores of central Honshu. Local limestone is used but iron-ore and coal are brought by giant bulk-carriers from Australia, Brazil and North America. The plant is situated in the middle of the nation's main industrial region which is linked by cheap water transport to local customers and to worldwide export markets.

Now try Exercise 10.

Service industry

Fig. 31 shows the importance of service jobs in Britain. Most people work in services such as retailing, banking and transport, which help primary and manufacturing industries to operate more efficiently. Services such as health centres, schools, restaurants and cinemas, swimming-baths and sports centres are needed wherever people settle in large numbers. Most services, therefore, are based near to their customers – in towns and cities. Eight out of ten people in Britain live in towns. Six out of ten of the nation's work-force are employed in services. Fig. 31 shows that even higher figures are recorded for places such as Cheltenham and Bournemouth.

Service occupations may be divided into four broad groups:
1. *Distribution* includes all occupations in transport and trade. Raw materials must be transported to factories and their products delivered to

customers. Wholesalers buy in bulk from manufacturers and distribute smaller consignments to retail shops. In a village where people grow most of their own food and have low incomes, the

Fig. 31 The percentage of the total work-force employed in services and manufacturing

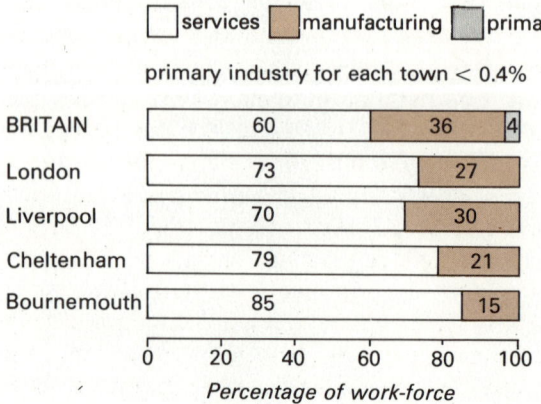

primary industry for each town < 0.4%

	services	manufacturing	primary
BRITAIN	60	36	4
London	73	27	
Liverpool	70	30	
Cheltenham	79	21	
Bournemouth	85	15	

Percentage of work-force

Fig. 32 The 'City' of London, Britain's financial centre

number of shops will be small. In a prosperous modern town, people with high incomes are able to buy a wide range of goods and services. For example, car-owners need garages for repairs, petrol stations for fuel, local offices for licences and insurance, and highway engineers and builders to keep roads in good repair. Each citizen becomes dependent upon the services provided by other people. One person's spending is another's income.

2. *Financial services* provide increasing numbers of jobs in towns. The traditional role of the money-lender was common in ancient times. Today private individuals, traders and industry seek a range of services offered by banks, accountants, insurance companies and building societies, post offices and social security departments. Many of these jobs involve personal contact with the customer and cannot easily be carried out by machine. Strong links between different financial operations such as banking, insurance and accountancy bring them together in central areas of cities and towns. An outstanding example is seen in Fig. 32. Here the narrow streets and high-rise offices alongside the Bank of England and the Stock Exchange comprise Britain's financial centre known as the 'City'.

3. *Administration* includes Central and Local Government office jobs. Thousands of civil servants are employed to assess and collect taxes. The growth of the 'welfare state' has been a notable feature of the twentieth century in Britain. Social services provide unemployment benefit and pensions, and help with problems concerning housing, health and education. Town-planning must also be carefully monitored, as must the country's defence and policing.

4. *Personal services* include those carried out by many professional people such as lawyers, doctors, dentists, actors and entertainers, hairdressers and chiropodists, and finally, undertakers. Only in towns are there sufficient numbers of customers to justify such a wide range of services.

23

Service industry: distribution

Supermarkets

Most families buy their weekly groceries from a local supermarket. Supermarkets were first introduced in the United States in the 1930s and were established in Britain after 1950. Stores such as Finefare, Safeway, Tesco and Sainsbury's have become familiar household names and have gained a large share of the grocery market. Unlike the small local grocer, supermarkets depend upon a large threshold population prepared to travel further for low-order goods. Their success depends upon their ability to attract large numbers of customers who buy in bulk each week. This is achieved by offering customers competitive prices in a wide range of grocery and household items. Although only a small profit is made on each item, high sales mean that the total amount of profit made is high (see Fig. 33). This high turnover enables supermarkets to place very large orders with manufacturers. They in turn quote low prices for bulk orders. Supermarkets cut staffing costs by operating self-service displays with cash check-outs. Their success may be judged by the fact that most neighbourhood shopping centres have at least one supermarket.

The hypermarket revolution

Large-scale shopping has now been taken a step further. Whereas supermarkets cover a floor space of about 250 square metres, a modern *hypermarket* such as the one at Eastleigh in Hampshire (Fig. 34) has a ground floor area of 11 200 square metres of which 7500 square metres is devoted to selling. The total site of the Eastleigh hypermarket, including car-parks, is 52 500 square metres – an area sufficient for more than seven football pitches.

Fig. 34 Cash checkouts at Eastleigh hypermarket

Hypermarkets offer 'one-stop' shopping for a wide variety of low- and middle-order goods such as food and drink, books and bikes, wallpaper and shoes, garden spades and even kitchen sinks. At Eastleigh, somewhat less than half the floor space is devoted to groceries. The hypermarket is really a huge selling factory. The advantages of its large scale are passed on to customers in the form of low prices and variety of goods. The extensive area of land it needs, however, can only be obtained cheaply (and planning permission given) if it is situated outside the built-up area of the town. The success of the hypermarket depends on its customers owning cars. At Eastleigh, 19 out of 20 customers travel to the store by car. They shop, typically, once or twice a month and buy bulk quantities of groceries. These are largely in canned or frozen form, and so can be kept for weeks. Hypermarkets achieve great savings in costs by ordering large consignments direct from

Fig. 33 Large- and small-scale profits

VILLAGE GENERAL STORE		SUPERMARKET	
DAILY SALE		DAILY SALE	
Canned beans (from wholesaler)	*per tin*	Canned beans (from manufacturer)	*per tin*
cost price	25p	cost price	23p
selling price	30p	selling price	25p
profit	5p	profit	2p
total sales	10 tins	total sales	1000 tins
total profit	50p	total profit	£20

the manufacturer, so that goods can be offered up to 20% cheaper than at smaller, rival stores. A recent survey of the Eastleigh hypermarket shows that it attracts over 30 000 shopping visits each week and that on average expenditure per trip is £20. Thus, total sales are about £30 million per year.

Future patterns of shopping

Because they are situated outside the built-up area, hypermarkets may help to reduce traffic congestion in city centres. Town centre and sub-urban shops may well suffer a loss of trade as more hypermarkets are built. This could lead to high-order retail shops moving out of the city centre to sites nearer to the hypermarkets themselves – a trend seen in the United States. For those without

Exercise 11 Hypermarket shopping

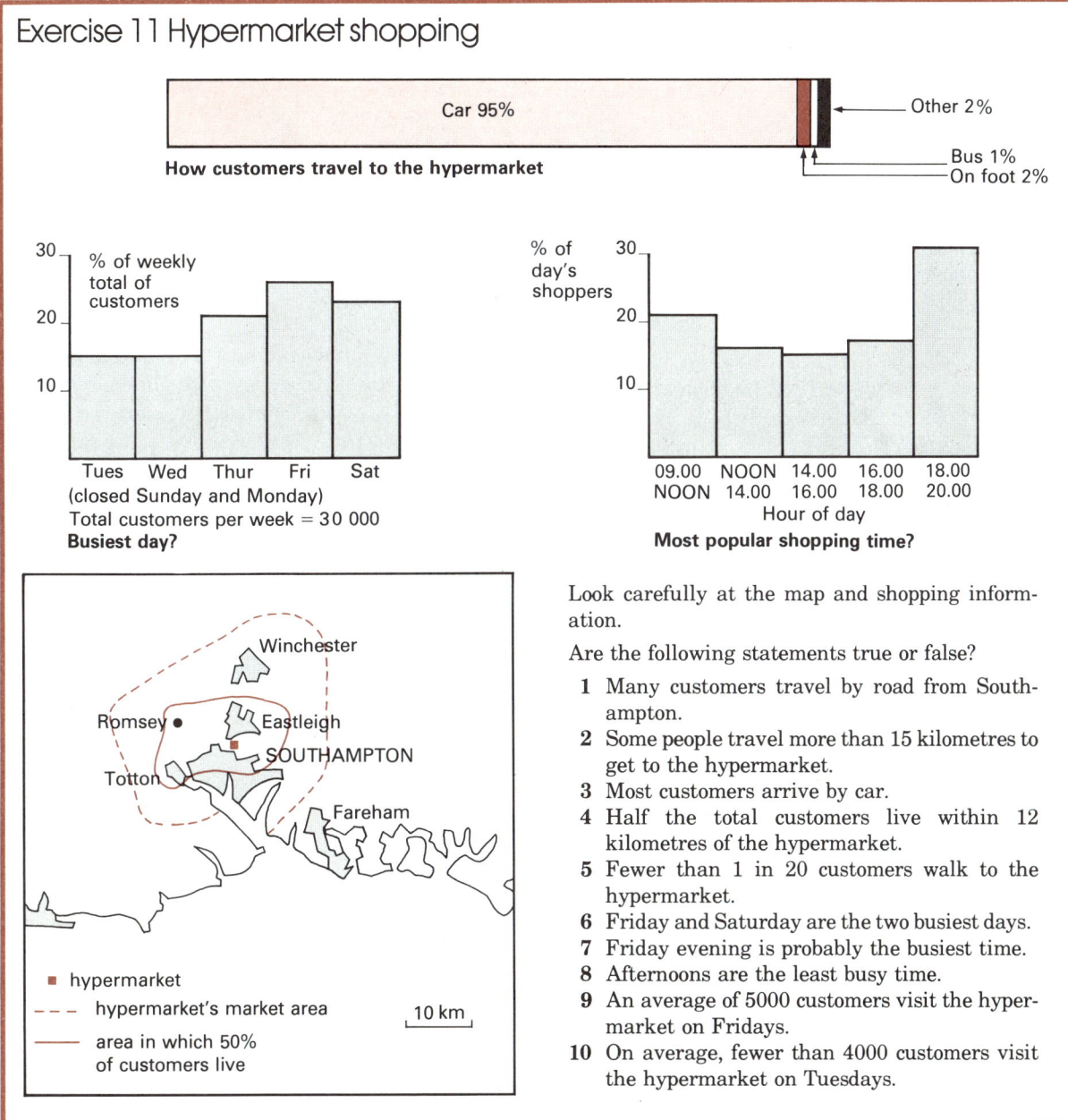

How customers travel to the hypermarket

Car 95%
Other 2%
Bus 1%
On foot 2%

% of weekly total of customers

Tues Wed Thur Fri Sat
(closed Sunday and Monday)
Total customers per week = 30 000
Busiest day?

% of day's shoppers

09.00 NOON 14.00 16.00 18.00
NOON 14.00 16.00 18.00 20.00
Hour of day
Most popular shopping time?

Winchester
Romsey
Eastleigh
SOUTHAMPTON
Totton
Fareham

■ hypermarket
--- hypermarket's market area
— area in which 50% of customers live

10 km

Look carefully at the map and shopping information.

Are the following statements true or false?

1 Many customers travel by road from Southampton.
2 Some people travel more than 15 kilometres to get to the hypermarket.
3 Most customers arrive by car.
4 Half the total customers live within 12 kilometres of the hypermarket.
5 Fewer than 1 in 20 customers walk to the hypermarket.
6 Friday and Saturday are the two busiest days.
7 Friday evening is probably the busiest time.
8 Afternoons are the least busy time.
9 An average of 5000 customers visit the hypermarket on Fridays.
10 On average, fewer than 4000 customers visit the hypermarket on Tuesdays.

cars, the increasing success of hypermarkets may bring hardship. Neighbourhood shops may go out of business or charge higher prices to survive. This would particularly affect people on low incomes who cannot afford cars and people who do not want cars. In the long term, if smaller shops do go out of business, leaving hypermarkets with less competition to face, the hypermarkets would find themselves in a strong position to raise prices without losing customers.

Now try Exercise 11.

The localisation of industry

Industries such as baking, brewing and printing benefit from a location close to their customers and are called *market-based* industries (see page 14). They are also examples of scattered or *dispersed* industry and are found in most towns. Some industries, however, are *localised,* that is, the majority of factories in the industry are situated in a small number of places. Fig. 35 shows the location of pottery manufacture in Britain. More than 75% of the nation's output comes from one small area mapped in Fig. 36. Other examples of localised industry are shipbuilding on Clydeside and Tyneside, glass manufacture at St. Helens, and the manufacture of high-quality steel in Sheffield. Although employment in these areas is no longer dominated by just one industry, cities such as Stoke-on-Trent and Newcastle upon Tyne remain heavily dependent upon a small range of products.

The reasons why a particular industry is localised are often historical. Iron was produced in the Sheffield district in the twelfth century. Local iron-ore, found in coal measures, was smelted in charcoal furnaces supplied from nearby forests. Grindstones were made from local sandstone; furnace bellows, forge hammers and grinding wheels were driven by water-power supplied from fast-flowing Pennine streams. Specialisation in cutlery developed at an early stage and although today iron is no longer smelted in the area, Sheffield supplies over 40% of the nation's high-grade steel products such as cutting instruments and machine tools. More than 1 in 5 of the city's work-force is employed in metal manufacture.

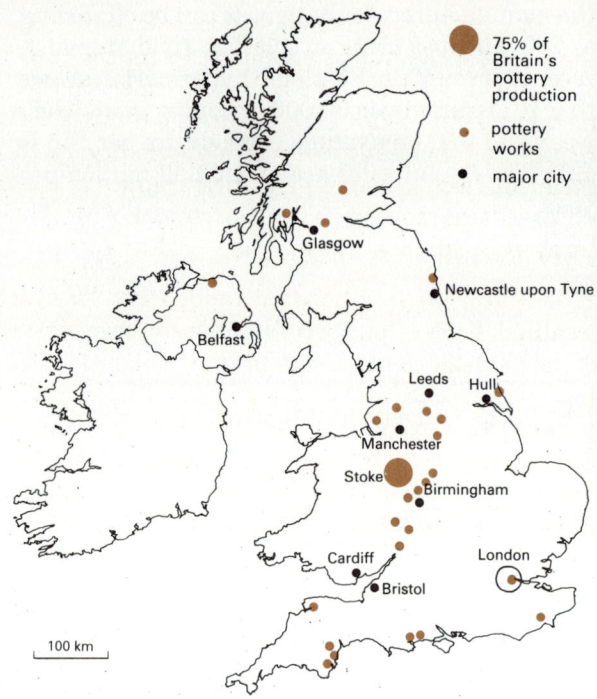

Fig. 35 The location of the pottery industry in Britain

The tendency for an industry to remain in the area where it first developed, even though the reasons for its early growth no longer apply, is called *industrial inertia.* It is explained by the growth of acquired advantages which the area offers a particular industry. The clustering of similar firms in a locality creates important benefits called *economies of concentration.* These are of four kinds.

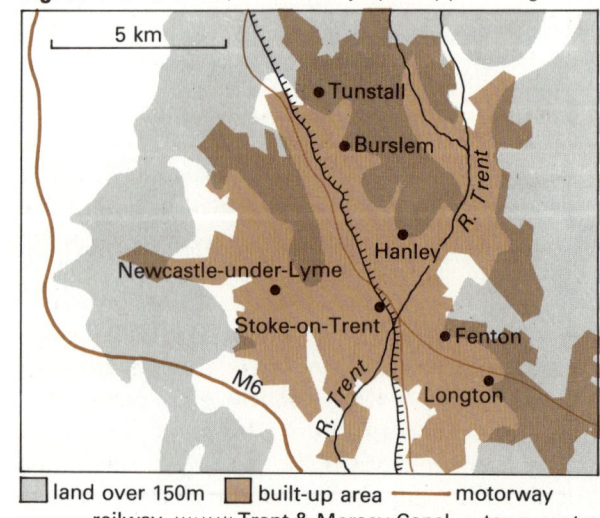

Fig. 36 'The Potteries', Britain's major pottery producing area

1. *Labour:* Where an industry is concentrated, the work-force develops special skills. In the case of Sheffield, the local labour force is expert in the metalworking and engineering techniques required by manufacturers of high-quality steel products. Local colleges gear their courses to the needs of local industry. The Department of Metallurgy at Sheffield University is world-renowned.

2. *Disintegration:* Where an industry is highly localised, firms often specialise in a single process of the industry. A good example may still be found in South Lancashire where the manufacture of cotton cloth is divided into separate processes such as spinning, weaving, dyeing and finishing. Here, many firms specialise in a single process, so they are able to produce on a larger scale and at lower cost. The separation of processes is called *disintegration*. In Sheffield, specialist firms concentrate on two main processes: the production of high-grade steel ingots from the electric arc furnace, and the finishing processes which transform the crude steel into, for example, cutlery, tools and engine parts.

3. *Co-operation:* The concentration of similar types of firm in an area encourages them to co-operate in research and development. Improved methods of production and new materials are therefore developed. For example, the Stoke-on-Trent pottery industry and the Lancashire textile industry have each established research institutes.

4. *Ancillary services:* Specialised industrial regions benefit from the emergence of subsidiary services which cater for the needs of the major firms. The concentration of industry may justify the building of motorways or improving railway services. Newly created industrial estates ensure that firms receive adequate supplies of gas, water and electricity. Banking and insurance facilities help to finance investment, provide credit and cover risks. Other specialists may develop in the area in order to supply component parts, machinery, or raw materials required by the major industry. By selling to several large manufacturers, subsidiary firms can produce on a large scale and at lower cost. A useful example from the Midlands, where the motor-vehicle assembly industry is localised, is provided by firms like Lucas and Smiths which supply most of the electrical equipment to the motor industry.

Although the concentration of an industry in an area may bring important benefits by reducing costs of production, specialisation may contribute to serious unemployment if the major industry becomes unprofitable. Lack of orders for the major industry results in a chain reaction in subsidiary firms. The absence of alternative employment in different industries increases the problem for the local work-force.

Changes in industrial location

Since 1920, several highly localised industries such as coal, cotton, iron and steel, and shipbuilding have experienced steady decline. These traditional industries were, and still tend to be, heavily concentrated in regions such as the Northwest, Clydeside, Tyneside and South Wales. They have contracted because former overseas customers have developed their own industries, or because markets have been lost to new industrial competitors such as Japan and West Germany. In some cases, technological progress has produced better substitutes (for example, plastics for metals). Improved methods of production have resulted in workers being replaced by machines.

Declining industries, however, present no serious problems if workers can move to jobs in growth industries such as telecommunications, electronics and distribution. Unfortunately, labour tends to be *immobile*. Workers are not easily re-trained for new jobs and not easily persuaded to move to another part of the country to find employment. Industries which have developed during the twentieth century (for example, motor vehicles, chemicals, electrical goods, aircraft, food processing) tend to be located in south-east England away from the traditional industrial areas. Firms have developed close to large urban populations such as London which provide both markets and labour supply. A major consequence of this changing distribution of industry has been persistently above-average unemployment in many regions outside south-east England.

Now try Exercise 12.

27

Exercise 12 Localised industry

1 Give three examples of localised industry and name an important centre in each case.
2 Give three natural advantages of the Sheffield area for the early development of iron smelting.
3 What is meant by industrial inertia?
4 Which of the following may be called economies of concentration:
  (a) skilled local labour;
  (b) local colleges running specialist courses;
  (c) local producers of components;
  (d) high unemployment in times of recession?
5 Explain the meaning of industrial disintegration.
6 Give three reasons for the contraction of British industries such as steel and shipbuilding.
7 Which of the following help to make labour immobile:
  (a) costs of retraining redundant workers;
  (b) family ties to the home area;
  (c) Government grants to assist in removals?
8 Name three expanding industries located mainly in the Midlands and south-east of England.
9 Which region of Britain experiences least unemployment?
10 Give two reasons why firms may be attracted to the London area.

People without work

In August 1982, more than 3¼ million people in the United Kingdom were unemployed; 1 in 7 or 14% of the total work-force. The steady upward trend in unemployment since 1964 is shown in Fig. 37. The official measure of unemployment is given as the number of people registered at local employment offices as out of work and looking for jobs on a certain day of the month. Official totals underestimate the true figure because some people who are not entitled to unemployment benefit have not registered as unemployed (for example, housewives who would like to return to work). Unemployment figures also hide the fact that many workers are on short or part-time work, or have taken early retirement. State financed employment schemes such as the Youth Opportunities Programme (YOP) and the Youth Training Scheme (YTS) also reduce the jobless total.

The causes of unemployment

During the normal course of events, there will always be some workers who find themselves unemployed. For example, people who lose or leave their jobs may well take some time to find alternative work. People with seasonal jobs such

Fig. 37 Unemployment in Britain from 1964 to 1982

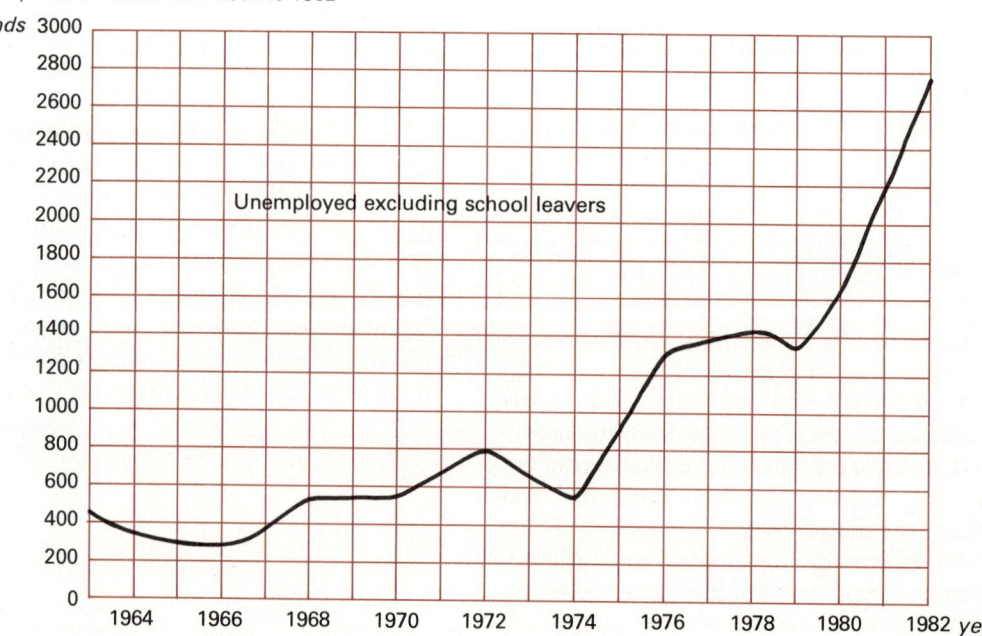

as seaside hotel workers are likely to be without a job for part of the year. This type of unemployment is called *frictional*.

An increasing number of workers, however, lose their jobs because of *structural* unemployment. This occurs when certain skills are no longer in demand. For example, containerisation has reduced the need for dockers to load and unload ships. The increasing use of the micro-processor (or silicon chip) is likely to reduce the demand for labour even further. This kind of unemployment in which workers are replaced by machines is also called *technological* unemployment. Structural unemployment in iron and steel, shipbuilding and textiles is largely due to severe foreign competition.

Recent unemployment figures for all regions of Britain have been made much worse by *cyclical* unemployment. Fig. 37 shows how the nation's economy seems to operate in a series of 'booms', when firms make profits and expand output, separated by periods of 'slump' when profits fall and unemployment rises. Since the huge increase in world oil prices in 1972-73, most industrialised nations, including Britain, have experienced the greatest slump (or *economic recession*) since the 1930s. In 1982, over 10 million people in the EEC were out of work – 1 in 10 of the work-force.

Now try Exercise 13.

Regional unemployment

Regional unemployment in Britain is mapped in Fig. 38. Note the location of regions with below average figures. Despite the low average for the South-East region, however, there are more unemployed workers in this densely populated area than any other– nearly 732 000, or about one quarter of the nation's total.

Fig. 39 shows that the regional imbalance in unemployment rates has persisted for many years. This has given rise to two serious consequences. Unemployed workers and empty factories represent a waste of economic resources which makes the nation poorer. The steady drift of workers seeking jobs in more prosperous regions leads to shortages of housing and overcrowding. Pressure is placed on roads, hospitals and schools

whilst similar facilities in other regions are under-used.

Regional policy in Britain

Government action to deal with regional unemployment began with the Special Areas Act, 1934, and since then a series of measures has been taken. The most important feature of Government policy has been 'to take work to the workers' rather than to help the unemployed to move to more prosperous regions. One reason for this is the *geographical immobility* of workers. An unemployed miner or steel-worker moving from South Wales to London faces many problems. The transfer of family and possessions is a costly operation. Apart from removal costs, there are expensive fees to be paid if selling and buying a house is involved. Furthermore, house prices and rents in south-east England are well above the national average. Many people do not relish the prospect of leaving friends, relatives and familiar surroundings to face a new life in a distant place. Families with young children are reluctant to move at crucial times in their education.

Exercise 13 People without work

Answer true or false.

1 By August 1982, more than three million were unemployed in Britain.
2 Unemployment has risen year by year in Britain since 1964.
3 Official unemployment figures include many workers who have not registered as unemployed.
4 Unemployed housewives form a major part of the official unemployed total.
5 Schemes such as YOP and YTS reduce unemployment amongst young people.
6 Structural unemployment occurs when workers are replaced by machines.
7 Typists who lose their jobs through the introduction of computerised word-processors are victims of technological unemployment.
8 The closure of motor-cycle factories causes cyclical unemployment.
9 'Booms' take place in periods of economic recession.
10 Cyclical unemployment affects all regions of the nation.

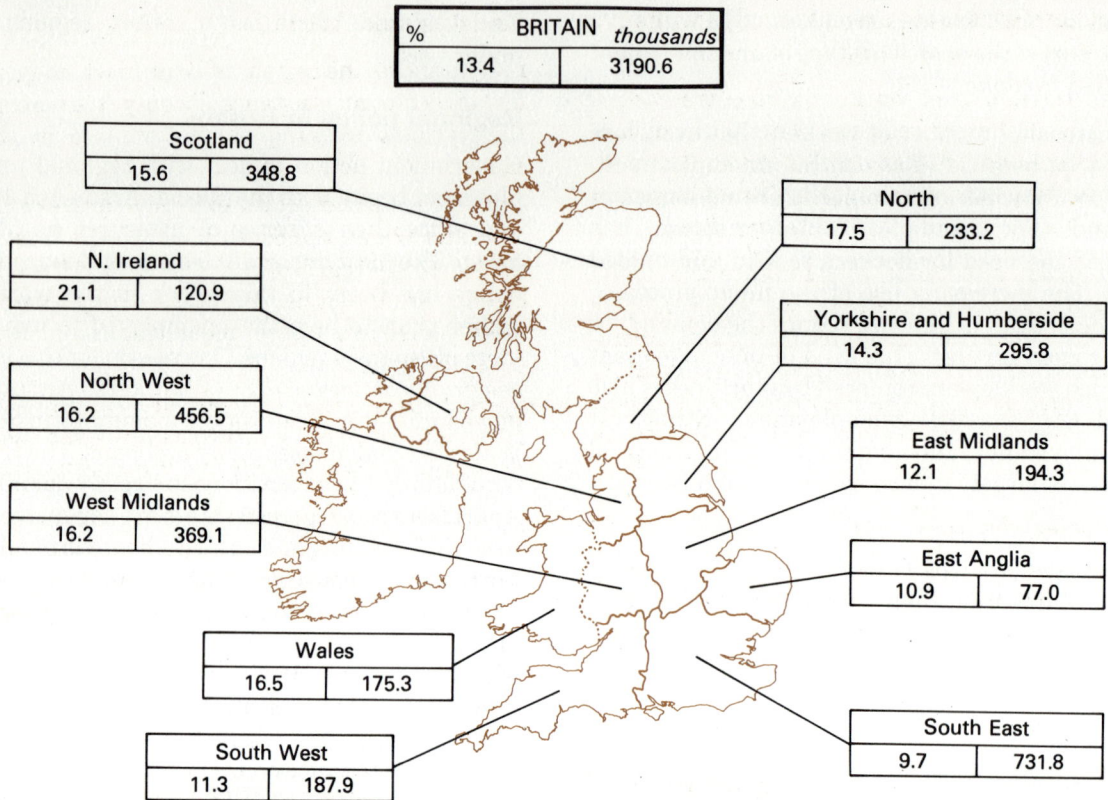

Fig. 38 Unemployment by regions (July 1982): percentage of work-force and number unemployed

Current measures to assist the regions are summarised below. They include, since 1973, financial help from EEC funds. They are of three kinds.

1. To improve the *infrastructure* of the declining areas by improving housing, roads, gas, water and electricity supplies, and by upgrading the provision of social amenities. This, it is hoped, will

Fig. 39 Regional unemployment rates (for selected years)

	1960	1967	1971	1976	1977	1978	1979	1980	July 1982
Standard regions	*Percentages*								
North	2.9	3.9	5.7	7.5	8.3	8.9	8.7	10.9	17.5
Yorkshire and Humberside	1.1	1.9	3.8	5.5	5.8	6.0	5.7	7.8	14.3
East Midlands	1.1	1.6	2.9	4.7	5.0	5.0	4.6	6.4	12.1
East Anglia	1.0	2.0	3.2	4.8	5.3	5.0	4.5	5.7	10.9
South East	1.0	1.6	2.0	4.2	4.5	4.2	3.7	4.8	9.7
South West	1.7	2.5	3.3	6.4	6.8	6.4	5.7	6.7	11.3
West Midlands	1.0	1.8	2.9	5.8	5.8	5.6	5.5	7.8	16.2
North West	1.9	2.3	3.9	7.0	7.4	7.5	7.1	9.3	16.2
Wales	2.7	4.0	4.4	7.3	8.0	8.3	7.9	10.3	16.5
Scotland	3.6	3.7	5.8	7.0	8.1	8.2	8.0	10.0	15.6
Northern Ireland	6.7	7.3	7.9	10.0	11.0	11.5	11.3	13.7	21.1
Britain	1.7	2.3	3.5	5.7	6.2	6.1	5.7	7.4	13.4

attract new firms to the areas and help to improve regional standards of living.

2. The retraining of workers whose traditional skills are no longer required. This will improve the *occupational mobility* of labour so that new firms moving into areas of high unemployment will be able to recruit the skills they need.

3. Financial assistance to stimulate industrial *expansion* and *diversification* (that is, a variety of different industries) in areas of highest unemployment, coupled with measures designed to restrict industrial developments in more prosperous or overcrowded regions such as the South East.

The Assisted Areas

Fig. 40 shows the regions of Britain which qualify for Government assistance. Twenty-five percent of the nation's working population live in these areas.

Development Areas cover extensive areas of Scotland and Wales, the South West, North West and the North; and parts of Humberside and the Corby district in the East Midlands. They comprise mainly districts where major traditional industries are in decline.

Special Development Areas are places within Development Areas with especially severe unemployment caused mainly by the closure of, for

Fig. 40 The Assisted Areas

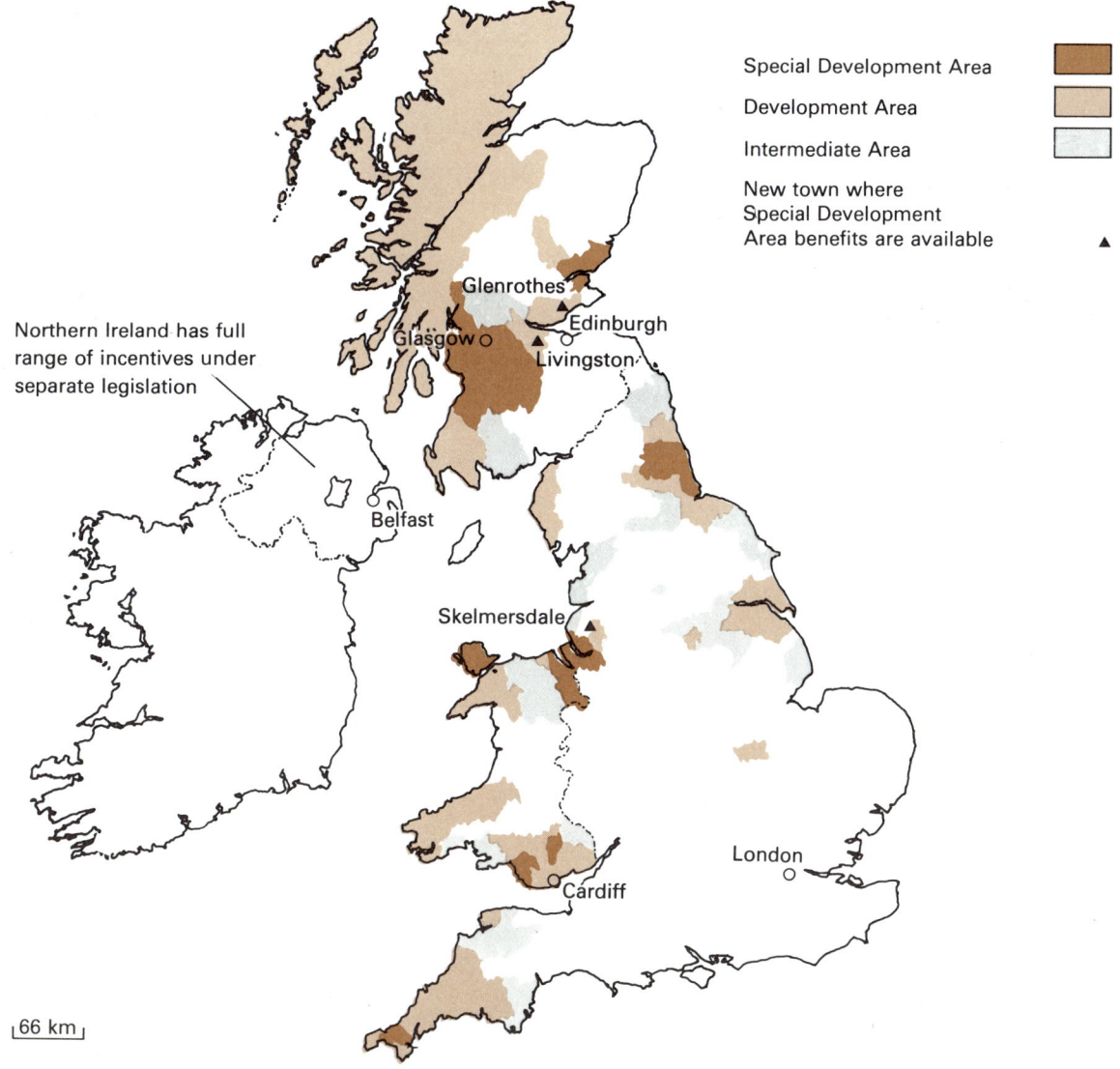

Special Development Area

Development Area

Intermediate Area

New town where Special Development Area benefits are available ▲

Northern Ireland has full range of incentives under separate legislation

Glenrothes ▲
Edinburgh
Glasgow ○ ▲ Livingston
Belfast
Skelmersdale ▲
London ○
Cardiff

66 km

Exercise 14 Regional unemployment

A Look closely at Figs. 38 and 39 before answering.

1 Which region has experienced the highest rate of unemployment since 1960?
2 Which region has experienced the lowest rate of unemployment since 1960?
3 In which region was more than 1 in 5 of the work-force unemployed in July 1982?
4 Which region has had fewer than 1 in 10 out of work since 1960?
5 In which region was the greatest number unemployed in July 1982?
6 In which region was the smallest number unemployed in July 1982?
7 Name five regions with above average unemployment rates in (a) 1960; (b) July 1982.

B

1 Describe two serious effects of regional unemployment.
2 Give three reasons to explain the geographical immobility of workers.
3 Explain what is meant by the infrastructure of a region.
4 Explain why Clydeside and Tyneside have been made into Special Development Areas.
5 Why has it been necessary to create Intermediate Areas?
6 Compare Fig. 40 with Fig. 38.
  (a) Which three regions contain no Assisted Areas?
  (b) In which region is the most extensive Special Development Area?

example, coal-mines, steelworks and shipyards. Clydeside, Tyneside, Merseyside and the ports of South Wales are included in this category.

Intermediate Areas, situated on the edges of Development Areas, are regions which receive Government·aid because they cannot otherwise offer the attractions of the prosperous regions nor the financial advantages of nearby Development Areas. Parts of North and South Wales, Devon and the North are included.

Now try Exercise 14.

Regional policy in South Wales

The South Wales region mapped in Fig. 41 first received Government aid under the Special Areas Act of 1934. The region was heavily dependent upon coal and steel. During the severe depression of the 1930s, the coal-mining valleys experienced unemployment rates of nearly 50%. The coal industry continues to decline. In 1956, 100 000 miners were employed in more than 100 pits; today fewer than 25 000 work in the 34 remaining

Fig. 41 Redevelopment in South Wales

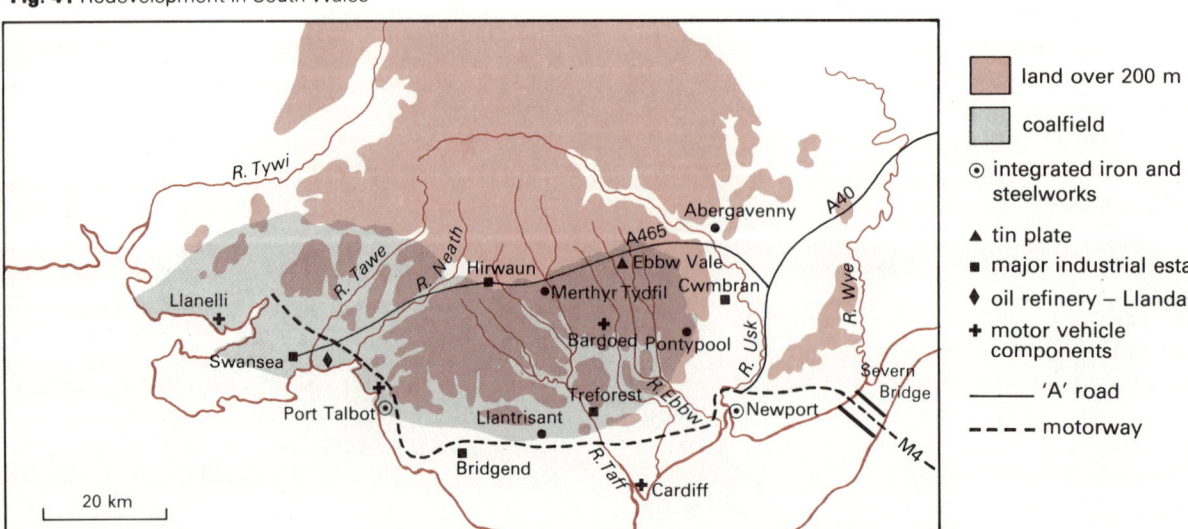

Legend:
- land over 200 m
- coalfield
- ⊙ integrated iron and steelworks
- ▲ tin plate
- ■ major industrial estate
- ◆ oil refinery – Llandarcy
- ✚ motor vehicle components
- —— 'A' road
- --- - motorway

20 km

mines and opencast operations. The steel industry, dominated by the giant integrated works of Port Talbot and Llanwern, is also threatened by further redundancies because of the reduced demand for steel.

Since 1934, Government policy has attempted to bring new growth industries to South Wales. In 1936, the Treforest Industrial Estate was opened at Pontypridd, 11 kilometres north of Cardiff. Today it provides employment for more than 10 000 in a variety of light industries which include car components, pharmaceuticals, electronic equipment, confectionery, leather goods, plastics and printing. Large industrial estates have also been established at Bridgend, Swansea and Cwmbran. Sites have been carefully chosen to be within easy reach of coal-mining valleys and redundant steel-workers. The valleys themselves are too narrow and remote for modern factories, so they have been located within easy access of new highways such as the M4 and the 'Heads of the Valleys Road', the A465. Improving the road network in South Wales is an important example of upgrading the infrastructure of the area in order to attract new firms.

Two very large employers brought to the region are ICI Nylon Spinners at Pontypool and Hoover Ltd. (domestic appliances) at Merthyr Tydfil. Oil refining at Llandarcy and chemical manufacture at Newport (Monsanto) have also expanded but these are *capital intensive* industries which employ relatively few workers. Many more jobs have been created by motor vehicle components factories in Cardiff, Bargoed, Swansea and Cwmbran. Valuable office work has been introduced by the Government. The new Royal Mint has been built at Llantrisant and the Driver and Vehicle Licensing Centre is at Swansea. Service employment of this type is especially valuable for a region lacking in jobs traditionally associated with female employment.

Opponents of regional policy argue that, in the long term, it does not help to reduce unemployment. Firms prevented from expanding in the South East may refuse to build new factories in areas where suitable labour is hard to find, and where markets are scattered and less profitable. Deliberate attempts to disperse an industry to several widely separated regions may lead to the loss of the economies of concentration. This increases costs of production, reduces profits and eventually threatens jobs. The dispersal of the motor vehicle industry away from its traditional home in the Midlands to sites on Merseyside and Renfrewshire in Scotland has been opposed by the industry. The closure of the Talbot plant at Linwood illustrates the difficulties of firms with factories located in high-cost regions faced with severe competition and recession. Some areas, it is said, are industrially exhausted and no amount of financial inducement will attract sufficient numbers of firms to restore employment to former levels. Many of the coal-mining valleys of South Wales are now being cleared of derelict pit-head installations and disused metal-works and the land reclaimed for forestry, farming and recreation.

Now try Exercise 15.

Exercise 15 South Wales

1 Which two industries dominated employment in South Wales in 1930?
2 How many coal-mining jobs have been lost since 1956 in South Wales?
3 What factor is threatening jobs in the steel industry?
4 Name four kinds of industry attracted to South Wales since 1945.
5 Why are industrial estates not established within the valleys?
6 Name one important way in which the infrastructure of South Wales has been developed since 1960.
7 Why do capital intensive industries *not* help to solve problems of unemployment?
8 Give one reason why service industries are especially helpful to combat unemployment in South Wales.
9 Explain two criticisms which have been made against the policy of assistance for South Wales.
10 Give two examples of successful Government aid to South Wales.

Localised unemployment

Although unemployment varies from one region to another (see Fig. 38), it also varies considerably within the same region. Fig. 42 shows pockets of very high unemployment in Consett, Ebbw Vale and Corby where recent steelworks closures have made a large proportion of the work-force redundant. The highest unemployment rates in Britain are now to be found in some of the inner suburbs of the largest cities. The relocation of factories, warehouses and offices from the city centre to spacious industrial estates on the outskirts has reduced job opportunities for residents in the inner city.

Fig. 43 shows the southern edge of Manchester's built-up area. In the foreground, a new housing estate has been built on land previously used for market gardening. The Roundthorn Industrial Estate can be seen in the middle distance. Many of the spacious single-floor warehouses and factories are occupied by firms which have moved from congested and cramped sites in the inner city. Good road links with nearby motorways such as

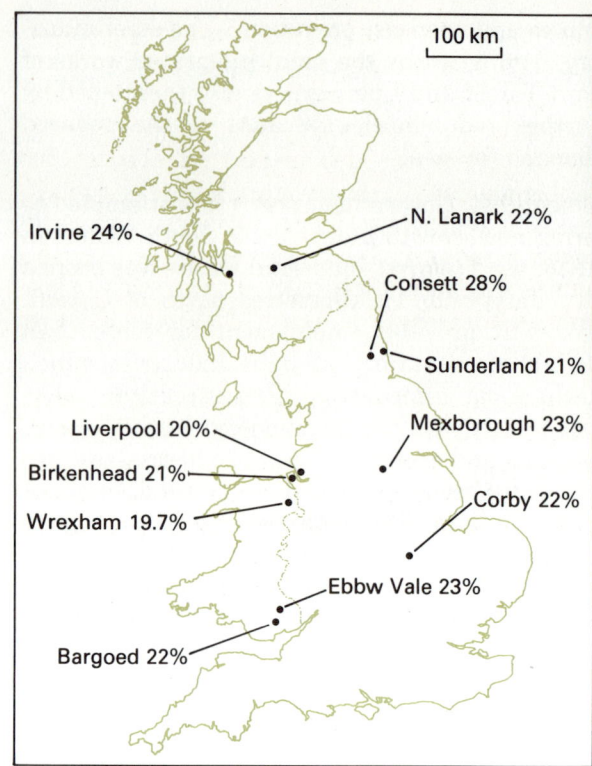

Fig. 42 Unemployment rates in selected towns (July 1982)

Fig. 43 The Roundthorn industrial estate, south of Manchester, near the M6 and M62 motorways

the M6 and M62 provide easy access for goods and workers. The estate gives employment to 2000 workers in baking, engineering, printing and photographic processing, and precision instruments. There are depots and warehouses for timber merchants, shoe wholesalers and electrical goods. The development of such estates helps to reduce congestion in city centres and can shorten journeys to work. Unfortunately, many of the jobs created are beyond the reach of those who live in the inner city.

Once inner suburbs begin to decline, they are no longer attractive to new sources of employment. High unemployment, especially amongst the young, and a decaying physical environment fosters vandalism, crime and loss of pride in the local community. In some districts, areas have been cleared of buildings, only to be left empty and desolate for years. Lack of public money for redevelopment and the reluctance of firms to move to such neighbourhoods hinders the introduction of much needed jobs.

Fig. 44 shows the rates of male unemployment in the inner suburbs of Liverpool. In April 1981, 18% of all men aged 16-64 were out of work in the Merseyside region. For Liverpool itself the figure was 22%; but for the inner suburbs close to the docks, rates varied between 34% and 38% – more than one in three of the male work-force. The proportion of unemployed men aged 16-24 is more than 50% in central districts. Fig. 45 shows the proportion of households in inner Liverpool with no car. Without such convenient means of transport, much of the work-force is immobile. Incomes are too low to travel far in search of work. Those fortunate enough to have jobs and prospects soon move to the outer suburbs in search of better housing and amenities, leaving behind a pool of ageing,

Fig. 44 The rates of male unemployment in Liverpool (April 1981)

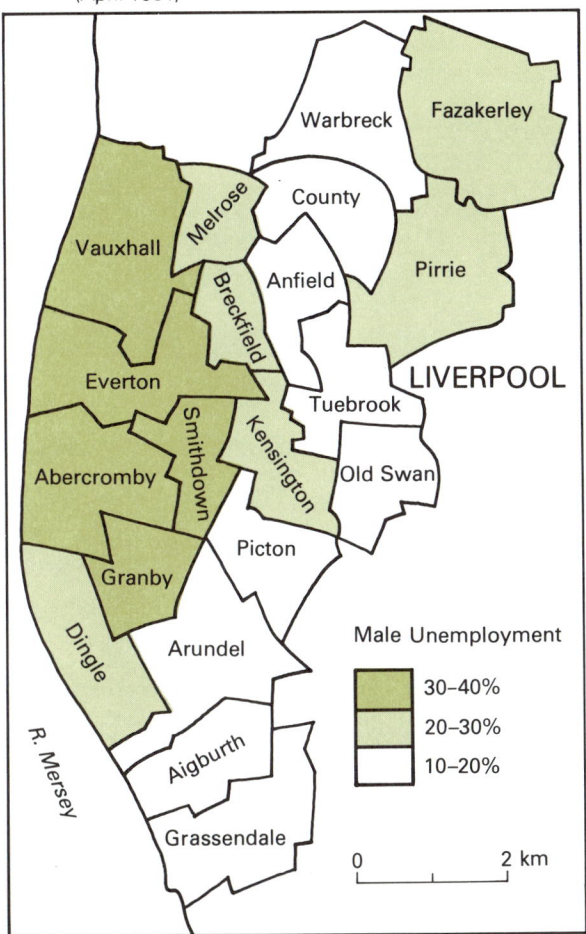

Fig. 45 The proportion of households in Liverpool with no car (April 1981)

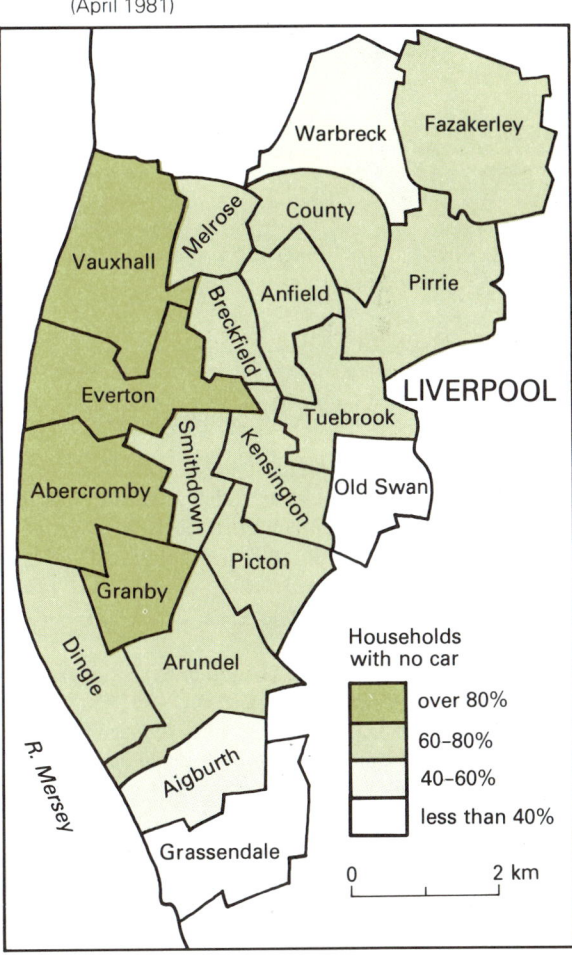

unskilled workers. A high proportion of inner city residents are poorly educated and lack the necessary skills needed by modern industry and commerce. In a world of computers and increasing automation, their chances of finding work are decreasing rapidly.

Fig. 46 The Enterprise Zones

100 km

Enterprise Zones

The emergence of localised areas of very high unemployment has resulted in the creation of *Enterprise Zones*. In September 1981, eleven such zones had been designated (Fig. 46). Under the Government's proposals, announced in 1980, in each zone of up to 2 square kilometres, firms will be granted freedom from paying rates for ten years. One hundred percent of the cost of building factories, warehouses and offices will be paid by public money. To persuade new firms to move into Enterprise Zones, planning regulations (other than health and safety) will be relaxed. In addition, firms qualify for the usual assistance available in Development Areas.

Such measures are designed to reduce unemployment totals in the inner cities of Glasgow, Belfast, Manchester, Newcastle, Liverpool and London; and in other unemployment-troubled areas such as Corby, Dudley and Wakefield. The levels of education and skill of the local work-force must be improved, however, if the jobs created are not to be taken by workers who live outside the inner city areas.

Now try Exercise 16.

Exercise 16 Localised unemployment

1 Answer true or false.
   (a) Steelworks closures have caused large-scale unemployment in Corby.
   (b) Inner cities have few problems of unemployment.
   (c) Industrial estates benefit from being situated close to city centres.
   (d) Industrial estates can help to reduce traffic congestion in towns.

2 Which of the following have contributed to high unemployment in the inner suburbs of Liverpool:
   (a) containerisation of docks;
   (b) dock closures;
   (c) demolition of factories;
   (d) low skills of residents;
   (e) Enterprise Zones;
   (f) closeness to the city centre;
   (g) low educational levels?

3 Describe three measures designed to attract firms to Enterprise Zones.

4 Which three of the following kinds of firms would be unlikely to move to an Enterprise Zone:
   (a) oil refining; (b) motor vehicle assembly; (c) bakery; (d) steel making; (e) printing; (f) brewing?

5 How many Enterprise Zones (Fig. 46) are designated for each of the following regions:
   (a) Scotland; (b) Wales; (c) the North; (d) the North West; (e) the South East; (f) East Anglia?

Chapter 2 The Power for Work

The demand for energy

Food crops receive energy from the sun. People and animals obtain energy from food. For many thousands of years, people depended upon their energy, together with that of domesticated animals, to provide the power for work. During the Industrial Revolution, however, other sources of energy began to be exploited. Machines were developed which relied on water-power and coal, rather than manpower, to operate. Factories were built to produce cloth and clothes, to process food and to manufacture goods. People moved to the towns to work in these factories and became dependent upon outside sources for food and clothing. This in turn increased the demand for mechanical, and eventually electrical, power to produce goods. The demand for power has grown throughout the nineteenth and twentieth centuries. Today we are becoming more and more dependent upon machines such as trains, planes, ships and road vehicles. These need huge quantities of oil-based fuels to power them. Meanwhile, electricity provides much of the power needed in our homes, factories, shops and offices. Electrical energy is, in turn, produced by burning vast quantities of coal, oil and gas, and is known as a *secondary* form of energy.

Fig. 47 shows the rapid increase in the world's use of energy. In 1980, the world consumed more than four times as much energy as in 1940. As world population increases, and nations try to improve living standards, so the demand for energy increases. However, the amount of coal, oil and gas in the Earth's crust is fixed and one day we shall have used it all. We must therefore try to find new sources of energy and learn to make better use of existing fuel supplies.

Early forms of power

The earliest form of power was manpower. The potter's wheel seen in Fig. 48 is powered by human muscle. For centuries the cart and the plough have been pulled by domesticated animals such as the ox, buffalo, donkey and, above all, the horse.

Fig. 47 World energy consumption

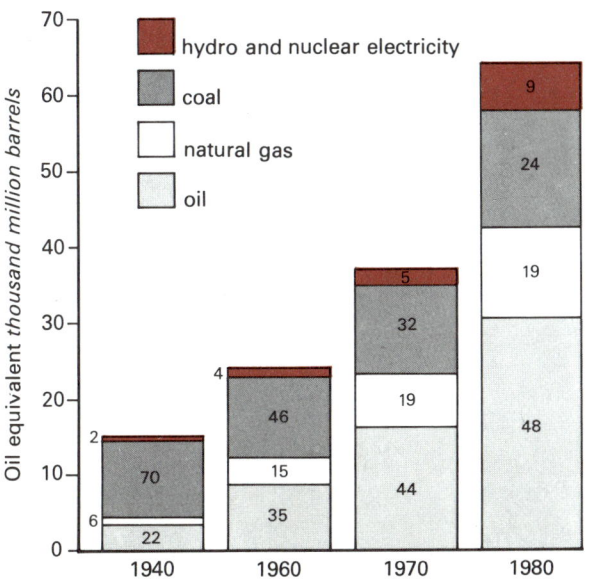

The figures show the percentage of total energy consumed.

Fig. 48 A hand-driven potter's wheel

Up to modern times wood has been burnt for warmth and cooking, for firing pots and making charcoal for smelting iron. Half the world's population still uses firewood for cooking and heating! Extensive areas of forest have disappeared in Africa and India because cutting has exceeded the natural regrowth.

During medieval times running water and wind were used to power flour-mills (Fig. 49). Wind-blown sailing-ships were not replaced by steamships until the late nineteenth century. Today, water power as the basis of hydroelectricity has become a major source of energy. Experiments to utilise the wind have proved less successful, although we may soon see sail-assisted supertankers as scientists and engineers look for alternative sources of energy.

Modern power

Since 1750, scientific discoveries and inventions have transformed the use of power. The Industrial Revolution was made possible by the invention of the coal-fired *steam-engine*. It was first used to pump water out of flooded coal-mines and was soon applied to new textile and iron-making machines. The new factories, supplied by steam railways, replaced the small-scale rural domestic system of making goods with large-scale urban industry. Coal is dirty, heavy and bulky, so high transport costs tied factories to the coalfields.

After 1880, however, the invention of the *turbine generator* led to the rapid development of electric power. Electricity is easily transmitted by overhead cables and is used to give light and heat, and to power electric motors. Factories are freed from the need to be near coal supplies.

Within the last hundred years, three new kinds of engine have transformed our methods of travel. All depend on oil to keep them going. The petrol-fuelled *internal combustion engine* appeared in the 1880s to drive the new 'horseless carriage' or motor car, and the *diesel engine,* invented in 1892, now powers the world's shipping. The *jet engine* (1939) burns paraffin to propel jumbo and supersonic aircraft around the globe.

◀ **Fig. 49** Preston Mill at East Linton in Lothian. This 17th century flour-mill is still in working order

Fig. 50 The 1320-MW Hinkley Point 'B' power-station near Bridgwater, Somerset

A recently-harnessed and increasingly widely-used source of energy is nuclear power. When used in a controlled fashion, it can be a valuable source of energy (Fig. 50). However, the atom bomb, and the destruction in 1945 of the Japanese cities of Hiroshima and Nagasaki, demonstrate the danger of nuclear power when used in war. It may nevertheless prove to be the most important source of energy in the twenty-first century when reserves of coal, oil and natural gas run out.

Now try Exercises 17 and 18.

Exercise 17 Sources of power

1 Rearrange these words to make a sentence: earliest provided muscles form the of power.

2 Describe one advantage of muscle power over mechanical power.

3 Why is electricity known as a *secondary* form of energy?

4 Give two reasons for the increase in world energy consumption.

5 Give two reasons to explain why an energy shortage is predicted.

6 Name four domesticated animals used to provide power.

7 Which two forms of power were used to drive flour-mills before steam-engines were introduced?

8 For which of the following has the steam-engine *not* been used:
textile machinery; water pumps; aircraft; railways; shipping?

9 Place the following in the correct historical order:
steam-engine, jet engine, petrol engine, diesel engine, nuclear power, sail, paddle.

10 Why were factories built near the coalfields in the nineteenth century?

11 Which form of power freed factories from the need to be near the coalfields?

12 Which of the following may be powered by electricity:
(a) road vehicle; (b) vacuum cleaner; (c) lawn-mower; (d) locomotive; (e) washing-machine; (f) refrigerator; (g) airliner?

13 Which of the following requires electricity to operate:
(a) radio; (b) television; (c) telephone; (d) motor car; (e) light bulb; (f) gas stove; (g) computer; (h) record-player; (i) video-recorder; (j) bicycle; (k) wheelbarrow?

14 Which of the following statements are true of electricity?
(a) It can be used to provide heat.
(b) It can be used to provide light.
(c) It can be made from coal or oil.
(d) It can be stored easily.
(e) It can be carried by overhead or underground cable.

15 Look carefully at Fig. 47 (world energy consumption).
a Which source of energy provided more than half the world's needs in 1940?
b Which source of energy provides nearly half the world's needs today?
c Has energy consumption in the last twenty years nearly doubled or more than doubled?

Exercise 18
The importance of energy

Complete the following passage using the ten correct words chosen from the list:

surplus, coal, poverty, mechanical, steam, hand-made, shortage, timber, population, Japan, Brazil, manufactured, soil, food, electricity, steam, earnings, central-heating, exercise, animal, machines

The world faces an energy_____. Reserves of _____, oil and gas are fixed and there is now a greater demand for power because world _____ and living standards are increasing. In poor countries such as India and_____, many people still use simple human and_____ power to farm the land, for transport and to make _____goods. The energy needed for such power is obtained from_____.

As living standards improve, factories use _____ to drive machines. Oil is needed for modern methods of transport. As_____increase, people use more energy for_____, motor cars, and household goods such as televisions, washing-machines, vacuum cleaners and refrigerators.

Fossil fuels

Today, most of Britain's power comes from coal and oil (Fig. 51). These are known as *fossil fuels* because they are formed from the remains of millions of trees and tiny creatures which lived millions of years ago. The total amount of fossil fuels in the Earth's crust is fixed so if we continue to extract them, reserves will eventually run out completely. They are therefore called *non-renewable* resources.

Coal

Britain produces 125 million tonnes of coal each year. This provides one-third of the nation's energy. Present reserves still in the ground will last for at least 300 years, and there are plans to increase production to 150 million tonnes per year by the year 2000. The importance of coal is likely to increase as oil becomes scarcer and more costly. Hydroelectricity and nuclear power are unlikely to supply more than a quarter of the nation's energy needs by the year 2000.

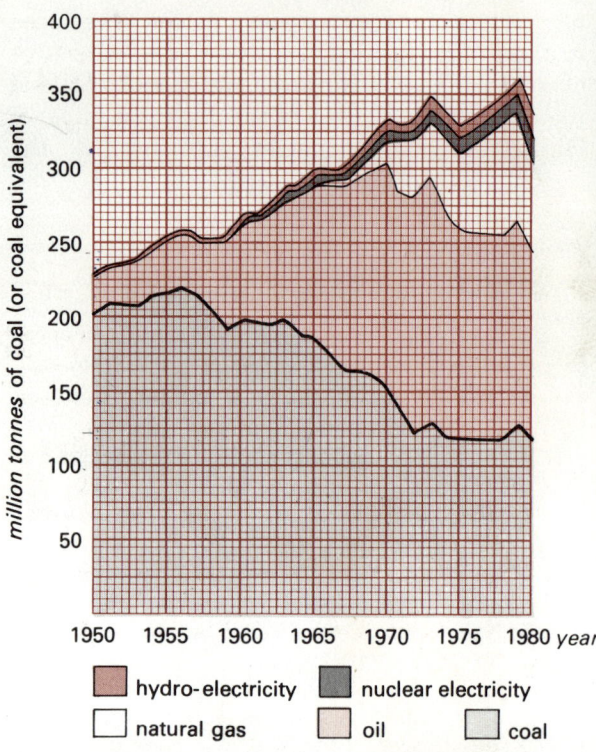

hydro-electricity		nuclear electricity
natural gas		oil
		coal

Fig. 51 Britain's energy consumption from 1950 to 1980

Formation of coal

The process of coal formation is shown in Fig. 52. The coal mined today was formed hundreds of millions of years ago when large areas of the Earth were covered in thick swampy forests. These were drowned when the sea-level rose, leaving thick layers of rotting vegetation. The vegetation was gradually covered by layers of sand and mud deposited by the sea. As these deposits thickened, new land was formed and new forests began to grow.

This process was repeated many times so that eventually, millions of years later, several layers of vegetation were created, each one separated by deposits of mud and sand. Over long periods of time, the layers of mud and sand were squeezed and hardened into *sedimentary rocks,* and the buried forest remains became seams of coal.

In some areas, where forest layers are more recent and have therefore been compressed much less, a softer kind of coal known as *lignite* is mined. This gives less heat than the hard types of coal. In wet marshland areas, rotting vegetation forms *peat*.

In Central Ireland peat is dried out and burnt to provide nearly one-third of Ireland's electricity.

Mining methods

Mining methods depend upon where the coal-seams are found. Fig. 53 shows three methods of extraction. Coal near the surface is removed by *opencast mining* (Fig. 53a). The topsoil is stripped away and set on one side. To uncover the coal, the rocks above it are removed by huge mechanical diggers. Powerful grabs dig out the coal which is then driven away by lorry. When all the coal has been extracted, the rock and soil are replaced and the area is reclaimed for farm land.

Where rivers have cut valleys into coal-bearing rocks, coal-seams may *outcrop* along the valley sides (Fig. 53b). Here, coal can be mined from tunnels dug into the seams. These tunnels, called *adits*, provide an easy method of mining and were used by many of the earliest miners.

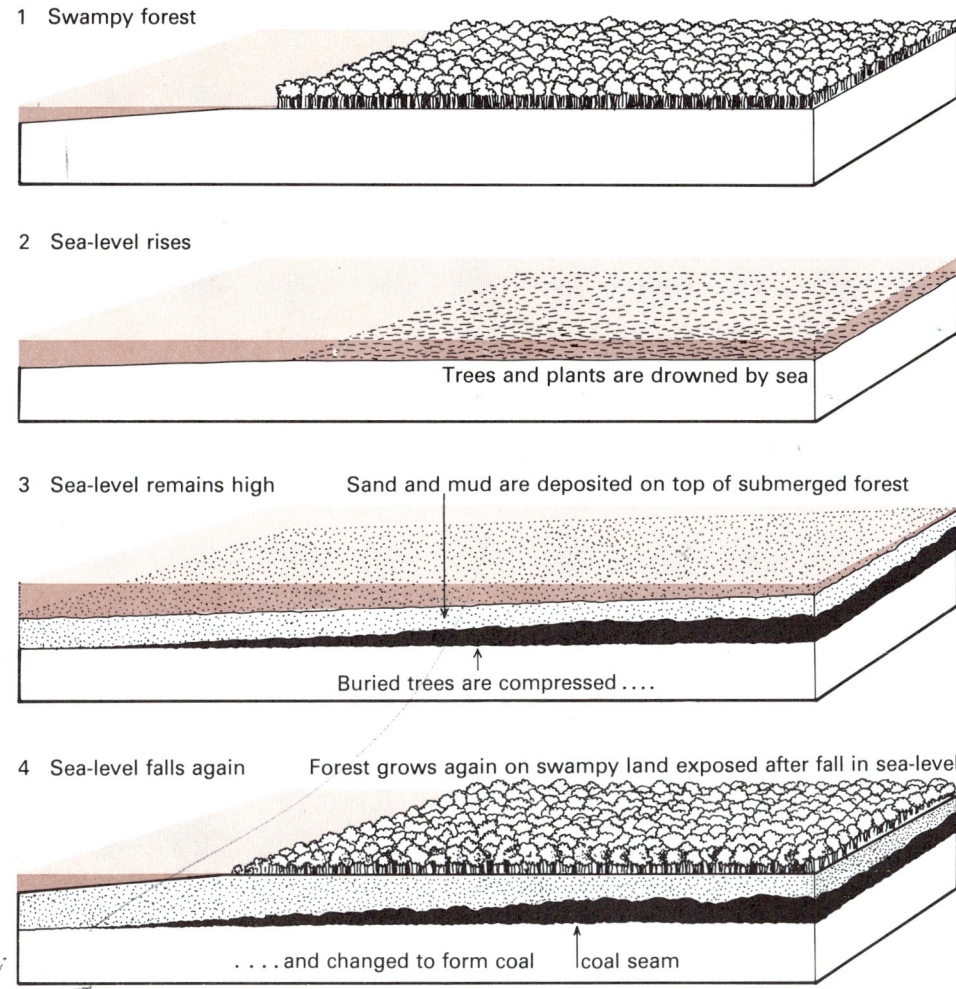

1 Swampy forest

2 Sea-level rises

Trees and plants are drowned by sea

3 Sea-level remains high   Sand and mud are deposited on top of submerged forest

Buried trees are compressed....

4 Sea-level falls again   Forest grows again on swampy land exposed after fall in sea-level

....and changed to form coal   coal seam

Fig. 52 The formation of coal

Today, the most common mining method is to sink *shafts* (Fig. 53c). These are cut down through the surface rocks to reach the deep seams. The coal is then removed from *galleries* which are cut into the seams. This is a costly and risky method. Miners face the dangers of flooding, roof collapse and explosions from underground gases. In older coalfield areas, the collapse of abandoned workings causes *subsidence* in the ground surface above (Fig. 53c). Expensive repairs are needed to prevent buildings and roads from sinking.

Modern machinery and methods mean that coal can now be extracted much more quickly. Hydraulic props hold up the roof whilst powerful shearers cut the coal from the coal-face before it is carried to the surface by machines.

Now try Exercises 19 and 20.

Fig. 53a Opencast mining

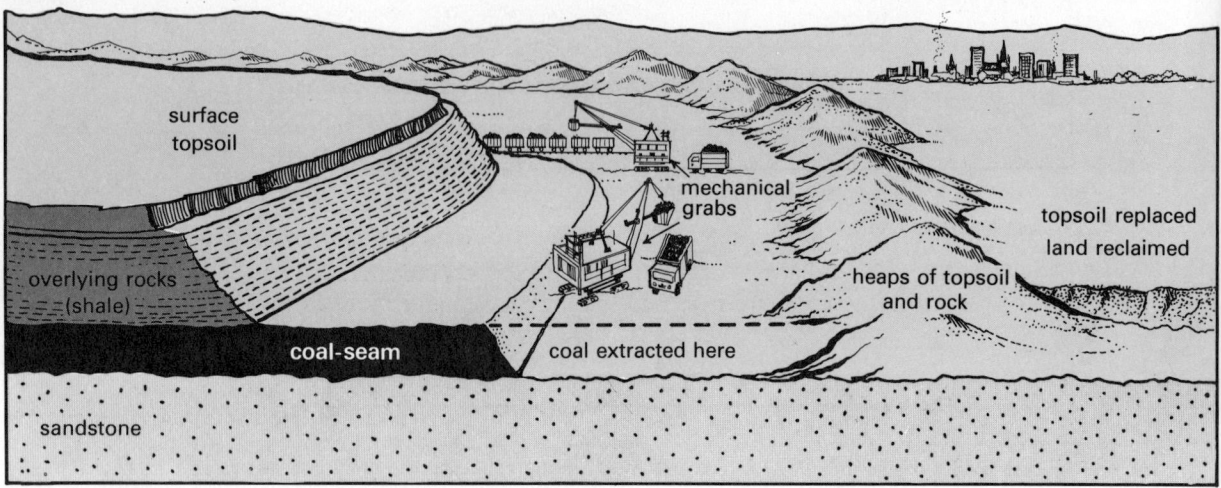

surface
topsoil

mechanical
grabs

topsoil replaced
land reclaimed

overlying rocks
(shale)

heaps of topsoil
and rock

coal-seam

coal extracted here

sandstone

Fig. 53b Adit mining

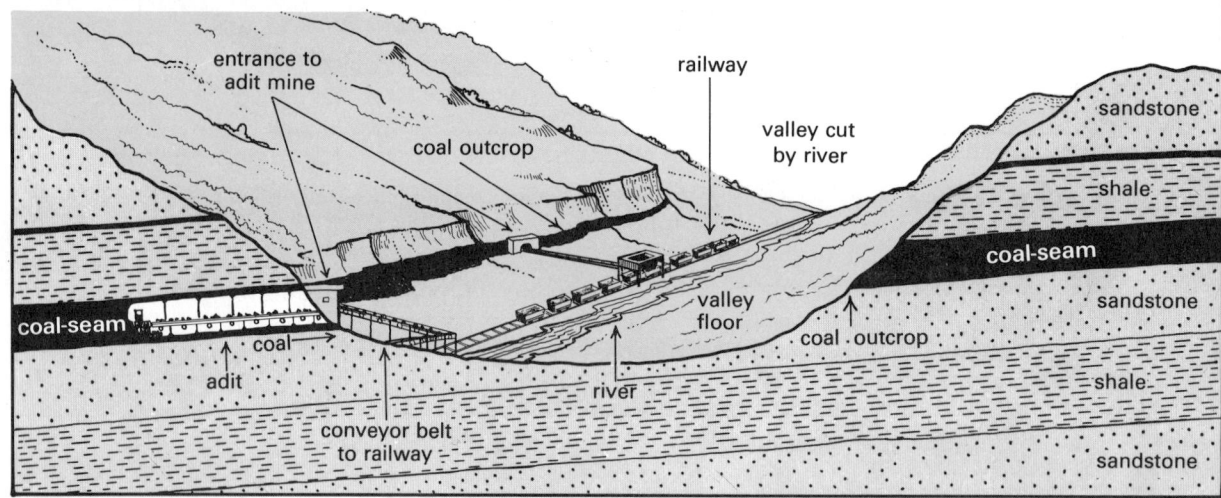

entrance to
adit mine

railway

coal outcrop

valley cut
by river

sandstone

shale

coal-seam

coal-seam

sandstone

coal

valley
floor

coal outcrop

adit

shale

conveyor belt
to railway

river

sandstone

Fig. 53c Shaft mining

railway

modern coal mine

road

mining subsidence

shafts

collapse of old workings
causes ground subsidence
at surface

sandstone

shale

coal-seam

gallery

coal-seam

sandstone

sandstone

shale

train

coal cut by machine at coal-face

Exercise 19 Fossil fuels

1 Which of the following are fossil fuels:
coal, oil, natural gas, timber, electricity?

2 Which of the following are renewable resources:
timber, solar power, tidal power, oil, animal power, HEP?

3 Place the following in order of age (youngest first): coal, peat, forest, lignite.

4 Name each of the following methods of coal-mining:
   (a) extracting coal from deep underground seams;
   (b) using a tunnel excavated into the side of a hill;
   (c) it takes farm land out of use for several years.

5 Which of the following are underground hazards faced by miners:
flooding; fire; explosion; roof collapse; frost; storms?

6 Which of the following terms are *not* connected with coal-mining:
shaft, outcrop, pithead, tanker, gallery, refinery, pit?

7 Give one reason why coal is likely to become an increasingly valuable source of energy.

8 Rearrange the following words to make sense:
non-renewable a coal is resource energy.

Exercise 20 Coal-mining

Look at the diagram carefully. The National Coal Board is considering new mining operations at the seven sites marked A-G.

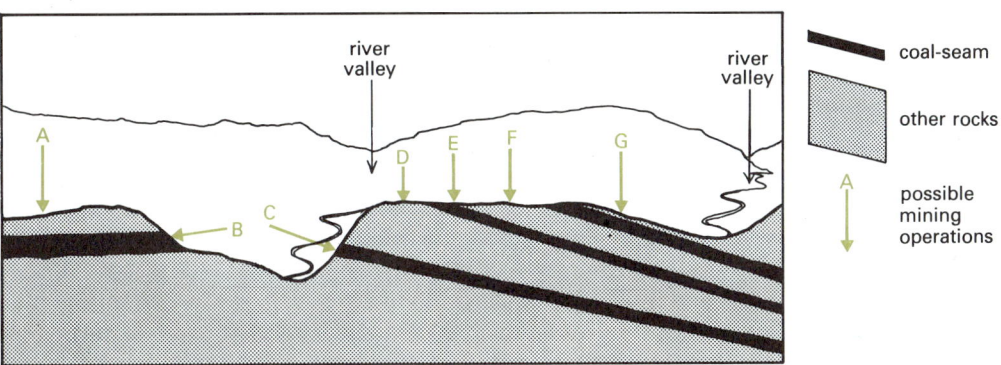

river valley

river valley

coal-seam

other rocks

possible mining operations

1 Which method of coal-mining will be used at each of the following sites:
A; B; C; D; G?

2 Give two reasons for sinking a shaft at F rather than at D.

3 Give two reasons why an adit mine is more likely at B than at C.

4 Give two reasons why opencast mining is more likely at G than at E.

5 From the diagram, say whether each of the following statements is true or false.
   (a) Mining from B may cause subsidence at A.
   (b) Mining from A may cause subsidence at C.
   (c) A shaft mine from D will be the deepest in the area.
   (d) Opencast mining at G will remove coal from three separate seams.
   (e) Opencast mining at G will cause serious subsidence at E.

Oil and natural gas

The world's oceans are swarming with tiny organisms called *plankton*. When the plankton die, the remains accumulate on the ocean bed and are buried in mud and sand. Eventually they decompose into tiny droplets of oil and bubbles of gas. Today's oil and gas began to form like this hundreds of millions of years ago. As oil and gas are both lighter than water, they gradually move up through the saturated porous rock. Eventually, they become trapped by an impermeable (waterproof) layer known as a *cap rock*. Where earth movements have folded these oil-bearing rocks into a dome, the oil and gas collects into a pool known as a *reservoir* (Fig. 54a). Fig. 54b shows a second type of reservoir. Here the Earth's crust has cracked and the rocks have moved along the line of the break or *fault*.

Once a suitable reservoir has been found, a well is drilled to tap the oil (Fig. 54a). The drilling bit is attached to lengths of steel pipe and driven into the ground from a platform known as a *drilling rig*. It is common for drilling to reach depths of 3 kilometres. Sometimes the first sign of a successful strike is a rush of natural gas to the surface. If the gas is present in large enough quantities it can be collected at the well-head for commercial use. Oil may rise to the surface under natural pressure; otherwise it is pumped out of the ground. At this stage it is a thick dark-coloured liquid called *crude oil*. This is processed into useful by-products such as petrol and paraffin at a *refinery* (Fig. 55).

Oil refining

Oil refining is a complicated and costly process used to split crude oil into its separate parts. These parts are known as *hydrocarbon fractions*. The operation is shown in Fig. 56. Crude oil is heated in a fractionating column to a temperature of 340°C. The oil begins to evaporate and the lightest hydrocarbons, such as petrol, rise up the column in the form of gas. As the gases rise they begin to cool and condense to liquid form at various levels in the tower. Here they are collected and piped to separate storage tanks. The heavier hydrocarbons, such as fuel oil and tar, remain in liquid form at the bottom of the tower. Notice that many oil products are used for different fuels for transport. Other refinery processes supply *petrochemicals* needed to make a wide range of familiar products such as plastics, detergents, nylon, paints and fertilisers.

North Sea oil and gas

The discovery of oil and natural gas in the rocks beneath the bed of the North Sea has provided a rich prize for the countries surrounding these stormy waters. Fig. 57 shows the location of the main fields. The first gas strike was made in 1965, 65 kilometres east of the Humber at the West Sole field. More than twenty other fields have since been found and now provide most of Britain's gas. Fig. 58 shows the remarkable increase in the

Fig. 54 Two kinds of oil reservoir

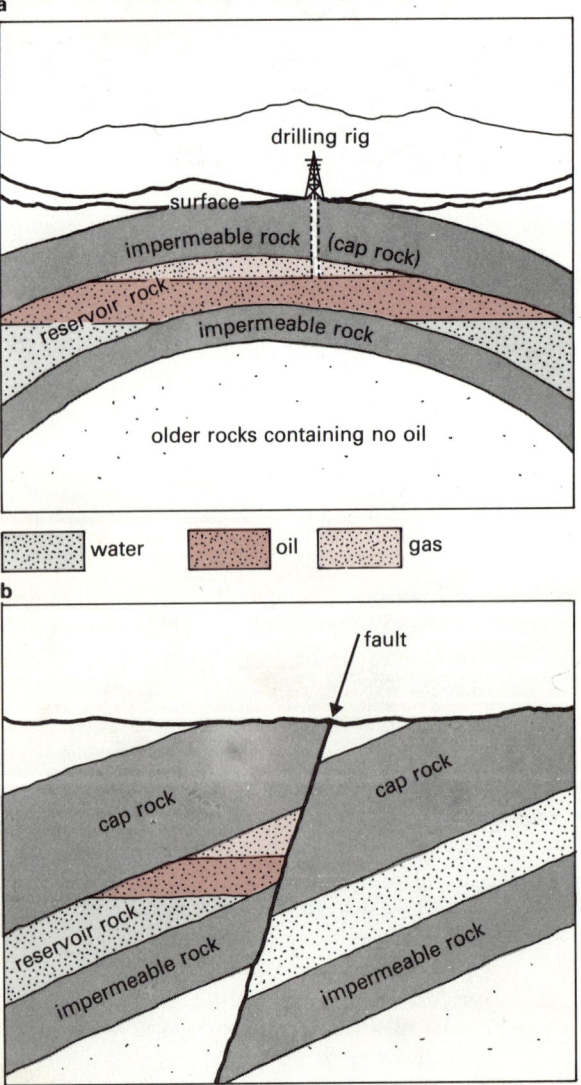

a

drilling rig

surface

impermeable rock   (cap rock)

reservoir rock

impermeable rock

older rocks containing no oil

water   oil   gas

b

fault

cap rock   cap rock

reservoir rock

impermeable rock   impermeable rock

Fig. 55 The crude oil distillation unit at Shellhaven on the Thames estuary

Fig. 56 The oil-refining process

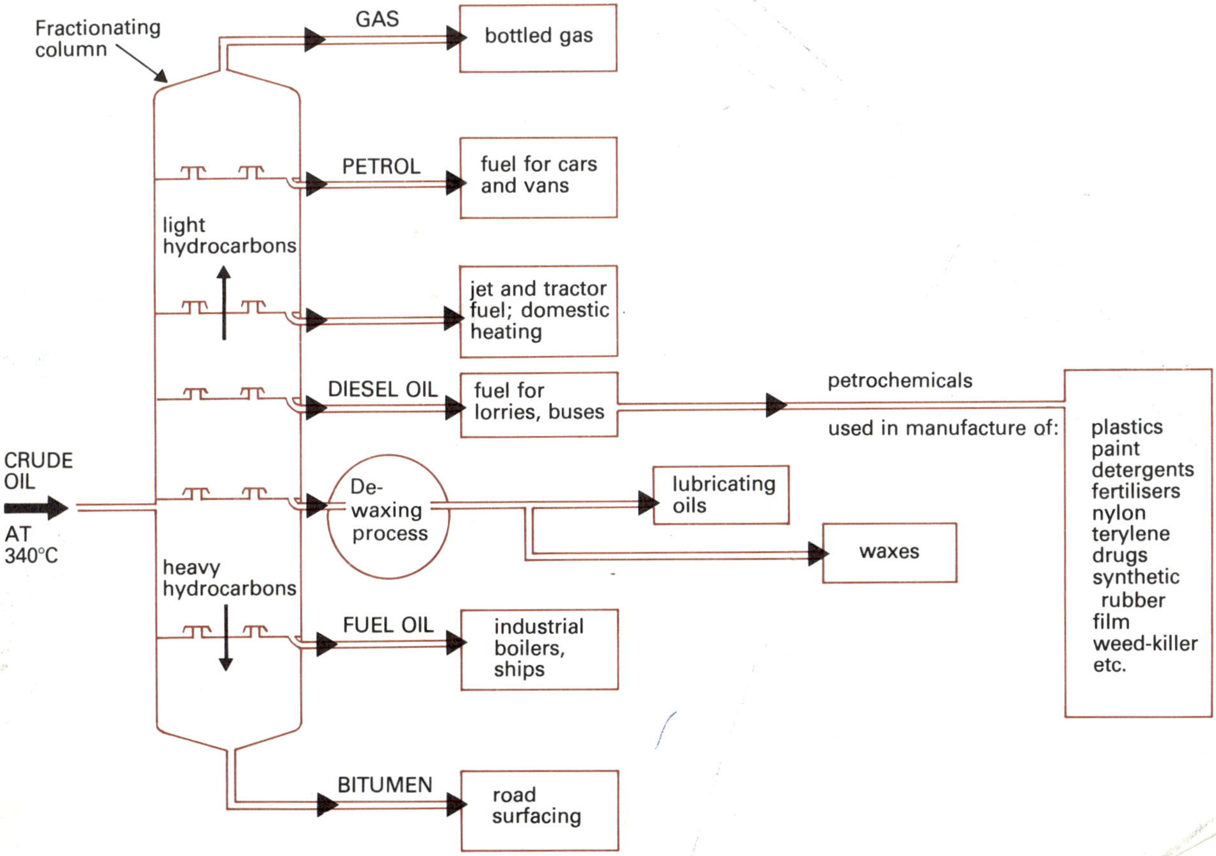

Fractionating column

GAS → bottled gas

PETROL → fuel for cars and vans

light hydrocarbons

jet and tractor fuel; domestic heating

DIESEL OIL → fuel for lorries, buses

CRUDE OIL AT 340°C

De-waxing process → lubricating oils

→ waxes

heavy hydrocarbons

FUEL OIL → industrial boilers, ships

BITUMEN → road surfacing

petrochemicals

used in manufacture of:

plastics
paint
detergents
fertilisers
nylon
terylene
drugs
synthetic rubber
film
weed-killer
etc.

amount of natural gas used, and the decline of coal gas. With reliable and convenient supplies delivered to most parts of the country by underground pipelines, natural gas is expected to provide one quarter of the nation's energy by the year 2000.

Fig. 57 North Sea oil and natural gas

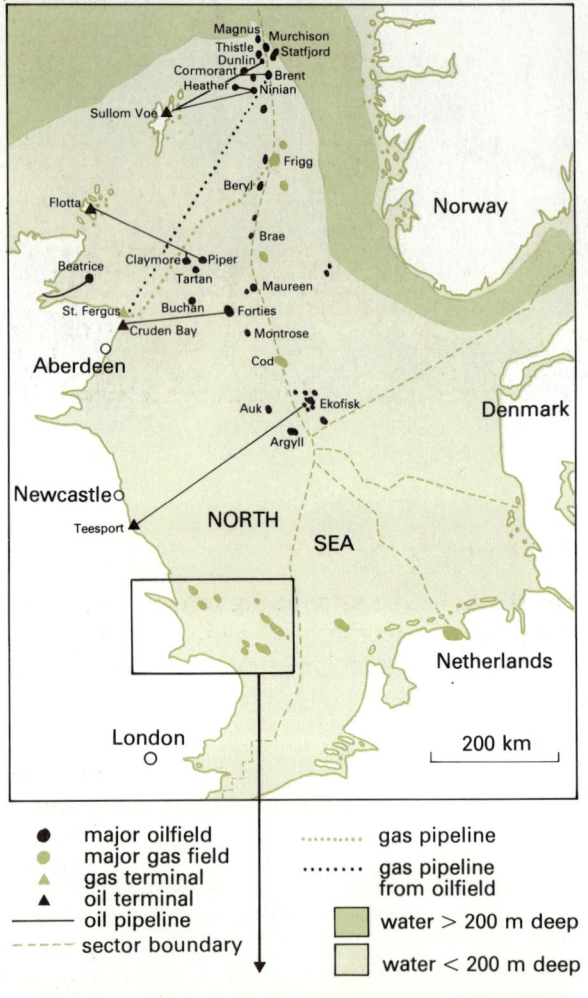

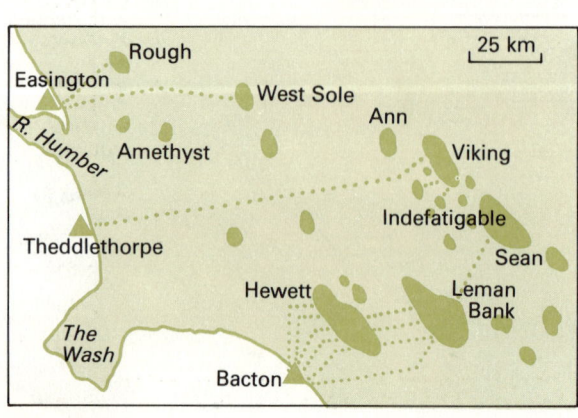

Gas consumption

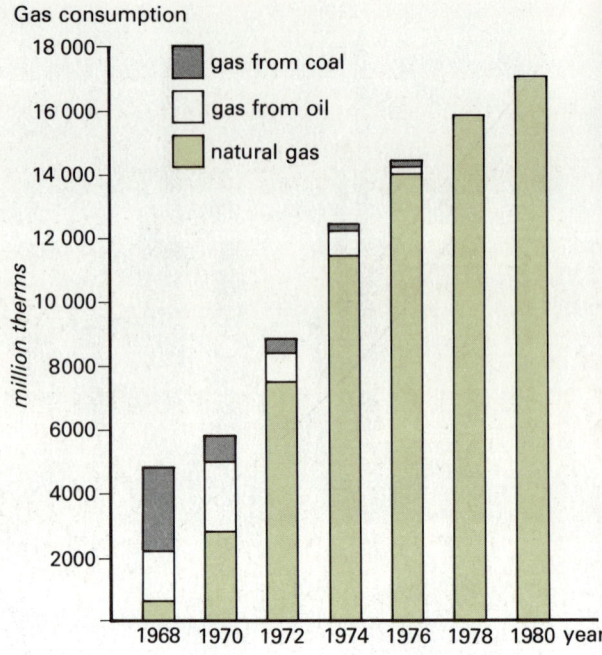

Fig. 58 Gas consumption in Britain from 1968 to 1980

Most oilfields are situated north of the latitude of Newcastle. The Forties is the largest of several major fields. Pipelines laid along the sea-bed bring the oil ashore at Teesport, Cruden Bay near Aberdeen, Flotta in the Orkneys, and Sullom Voe in the Shetland Islands (Fig. 57). The oil is of good lightweight quality and reserves should provide 60% of the nation's oil needs for at least 25 years, but Britain still needs to import heavier fuels from the Middle East. However, exports should earn valuable revenue for many years which hopefully will justify the difficulty in extracting the reserves.

North Sea oil and gas is extremely expensive to extract. Production costs are more than fifty times as much as Middle East oil. New drilling rigs specially designed for the North Sea cost many hundreds of millions of pounds to construct. Even greater sums are spent on coastal installations such as those at Sullom Voe (Fig. 59) which cost £1200 million to build. Fig. 60 shows the costs of developing a typical North Sea field. Such massive investment has only been made with the knowledge that reserves will last for many years and that fuel prices will continue to rise.

Fig. 59 The Sullom Voe oil terminal on Calback Ness, Shetland

Fig. 60 The costs of developing a North Sea oilfield

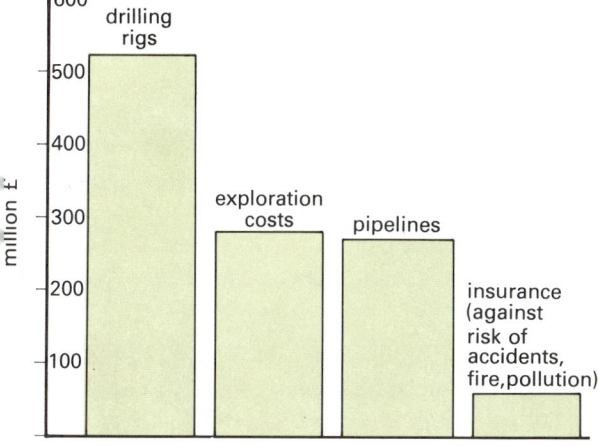

*The cost of building a terminal such as Sullom Voe is not included.
The operating cost each year is £50 million after oil starts to come ashore.*

Oil shortages

Fig. 61 shows the astonishing rate of increase in world oil consumption since 1950. This is largely due to its use as fuel for motor vehicles, diesel-

Fig. 61 World oil consumption from 1900 to 1980

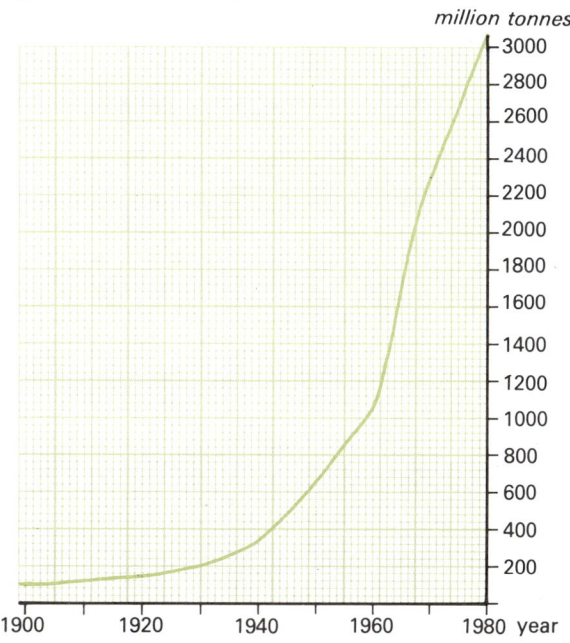

engined ships and paraffin-driven jet aircraft, and also its increasing use for petrochemicals. It is cleaner and easier to use than coal, and burns with 70% more heat than an equal weight of coal.

Sixty percent of the world's oil supplies are controlled by a small group of producer nations known as the Organisation of Petroleum Exporting Countries (OPEC). OPEC was formed in 1960 by Iran, Iraq, Kuwait, Saudi Arabia and Venezuela. Algeria, Ecuador, Gabon, Indonesia, Libya, Nigeria and the United Arab Emirates joined later. OPEC's objectives were to stabilise crude oil prices and unify export policies among its members. Until the 1970s, oil was relatively cheap, but between 1970 and 1980 its real price increased eight times. The oil-producing nations are anxious to preserve their reserves by controlling production and by maintaining prices. Because the world's industrial nations are dependent upon oil they will meet the extra cost.

New reserves recently found in the North Sea, Alaska and Peru will help to keep oil flowing, but the world's energy shortage is largely the result of dwindling supplies of oil and a steadily increasing demand for energy.

Now try Exercises 21 and 22.

Electricity

Electricity is made by a *generator*. The central shaft or *rotor* is made to spin inside a case of copper wires and this movement causes an electric current to flow. The rotor is turned by a *turbine* (Fig. 62). The turbine is based on the idea of a windmill whose blades rotate as the wind blows through. The blades of a turbine, however, are usually turned by jets of steam. Steam is produced by heating water in huge high-pressure boilers. Coal, gas or oil provides the fuel in a *thermal* power-station. In a *nuclear* power-station, the intense heat given off by *nuclear fission* is used to make steam (see page 57). In a *hydroelectric* power-station, falling water, rather than steam, turns the blades of the turbine. Waves and wind may also be used to generate electricity.

Electricity provides light and heat, drives motors and domestic appliances, and powers computers. It is a convenient, clean and flexible form of power, easily transmitted by a network of overhead transmission lines throughout Britain. This system, known as the National Supergrid (Fig. 63), is able to transfer current from areas of cheap supply such as the coalfields of the Midlands and South Yorkshire to the major towns of southern England.

Fig. 62 A simplified diagram of a coal-fired power-station

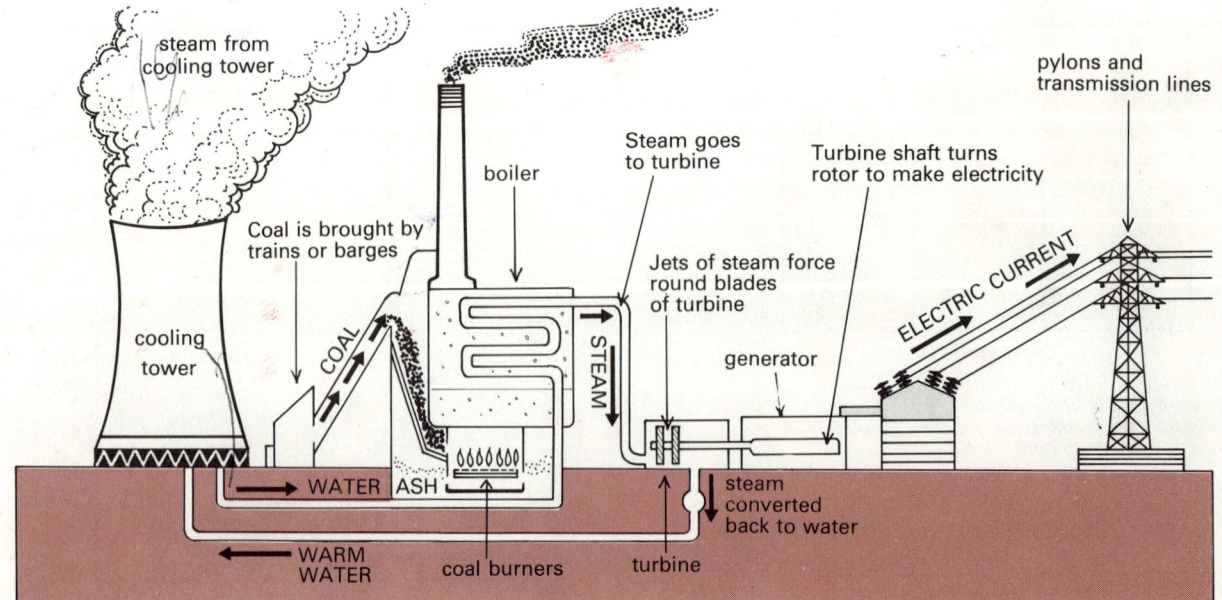

However, thermal electricity generated from fossil fuels (oil, natural gas, coal) is relatively expensive. It is a *secondary* form of power because fuel (known as *primary fuel*) is burnt to produce it. The process is not efficient because more than half the heat produced escapes in the generation process. Another disadvantage is that electricity cannot be stored in large quantities. At any minute of the day, a sufficient number of power-stations must be operating to supply the power needed. Unfortunately, the demand for electricity varies hour by hour, as can be seen in Fig. 64. Much less electricity is needed between midnight and 05.00. During the day there are 'peak' periods – for example, early evening when people return home; tea is cooked, lights and fires turned on and kettles

Exercise 21 Oil and natural gas

Choose the correct description.

1	crude oil	a crack in the Earth's crust
2	drilling rig	a layer of rock which covers a reservoir of oil and gas
3	oil refinery	a platform from which an oil-well is drilled
4	paraffin	a factory which splits crude oil into separate hydrocarbons
5	plankton	microscopic plants and animals which live in the sea
6	fault	chemical substances which make up crude oil
7	cap rock	a thick, dark-coloured liquid extracted from an oil-well
8	hydrocarbons	a hydrocarbon produced at an oil refinery

Exercise 22 Oil supplies

Answer true or false.

1 All North Sea oil belongs to the United Kingdom.
2 North Sea oil must be separated from sea-water in a refinery.
3 North Sea gas has replaced gas made from coal.
4 Most North Sea oilfields are situated south of the latitude of Newcastle.
5 The Forties is one of the largest North Sea gas fields.
6 North Sea oil deposits will make the United Kingdom self-sufficient in oil for 10 years.
7 Because North Sea oil is easily extracted, it is a cheap form of energy.
8 Sullom Voe is an important North Sea oilfield.
9 The world consumption of oil has more than doubled since 1970.
10 Oil is a renewable energy resource.
11 Fuel-saving cars will increase dependence upon oil.
12 Total world oil reserves cannot be increased.

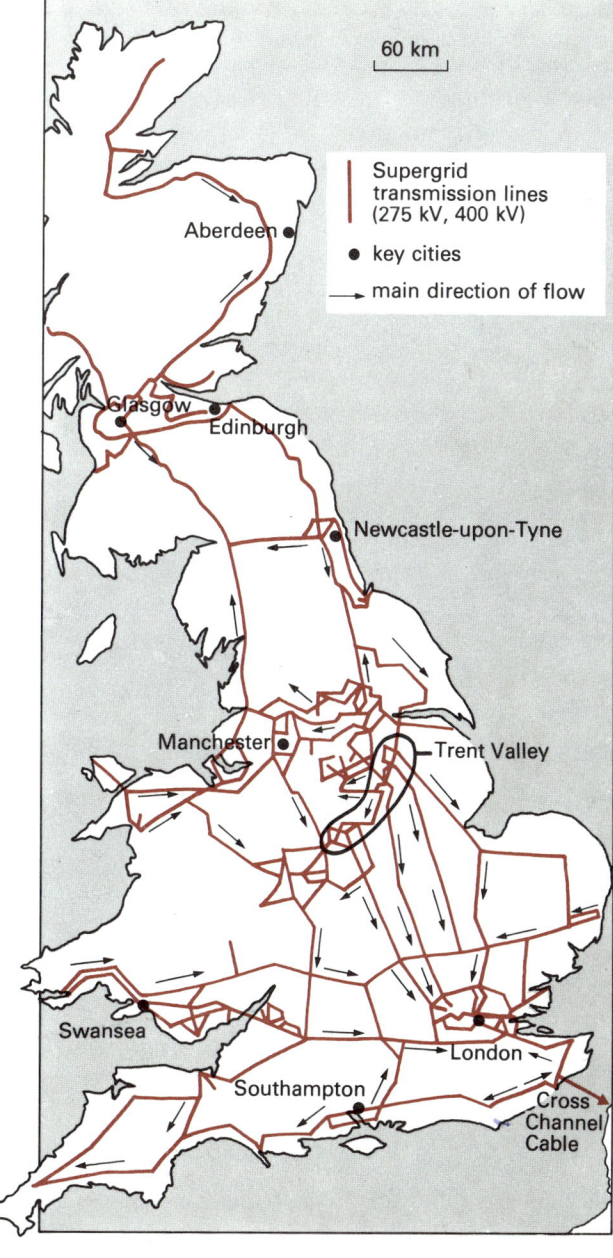

Fig. 63 The National Supergrid ▶

60 km

Supergrid transmission lines (275 kV, 400 kV)
● key cities
→ main direction of flow

Aberdeen
Glasgow ● Edinburgh
Newcastle-upon-Tyne
Manchester ● Trent Valley
Swansea
Southampton
London
Cross Channel Cable

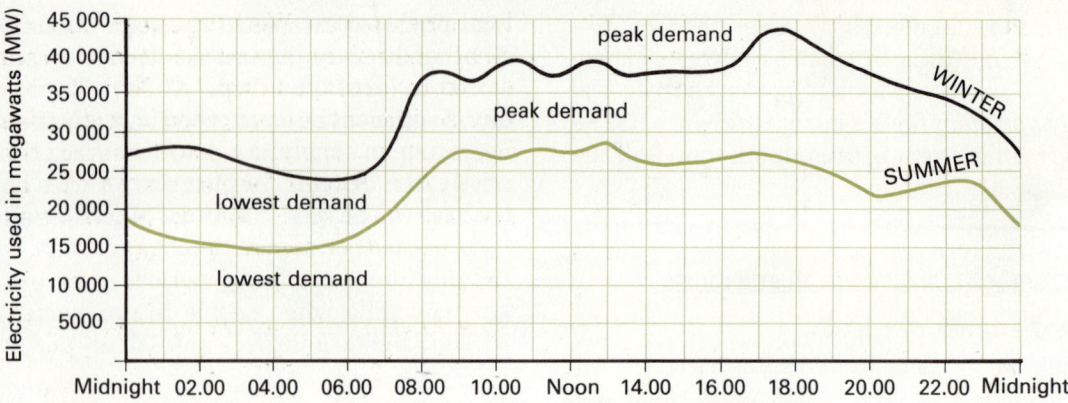

Fig. 64 Average daily electricity requirement during winter and summer

boiled. Popular television programmes cause peaks later in the evening. The total number of power-stations in the country must be able to produce enough electricity to meet peak demand. This peak demand may only last for an hour, so for the rest of the day many power-stations shut down. Because so many power-stations are idle for much of the day the cost of electricity is high.

Coal-fired stations

Coal-fired power-stations are situated near to fuel supplies because coal is heavy and bulky and, therefore, costly to transport. Vast quantities of water are needed for cooling purposes and specially treated mains water must be available to make steam. Wide areas of flat, firm ground are required to support the immense weight of the station, and for the large stockpiles of coal.

Fig. 65 The 2000-MW coal-fired West Burton power-station on the River Trent near Retford in Nottinghamshire

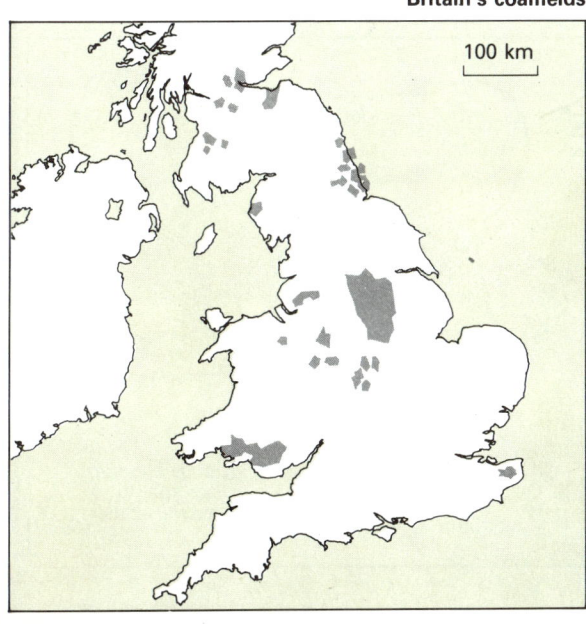

Thermal power-stations

100–1000	> 1000	
MW	MW	Fuel
·	●	coal
·	●	oil
○	○	gas

West Burton

Trent Valley

60 km

Britain's coalfields

100 km

Fig. 65 shows the West Burton power-station in Nottinghamshire. It produces 2000 megawatts of electricity – enough to supply 2 000 000 single-bar electric fires or the entire needs of a city the size of Birmingham. The site covers 450 hectares – an area large enough for more than 500 football pitches. It cost £150 million to build and took eight years from planning stage to completion in 1974.

West Burton burns on average just over 100 000 tonnes of coal each week. Supplies are carried by 'merry-go-round' trains which run daily from mines in Yorkshire and the east Midlands. Arriving at the power-station, the trains follow a loop line around the coal stockpiles. Within minutes, wagons have tipped their loads and departed for more. Each day the station produces 5000 tonnes of ash, some of which is used for making building blocks or used as a filler for reclaiming marshland and claypits.

Fig. 66 shows that West Burton is one of a cluster of similar power-stations built along the Trent Valley in the north Midlands. Over 85% of Britain's electricity is supplied by coal- and oil-fired power-stations like these. Notice how most of them are situated near coalfields – especially in Yorkshire and the east Midlands. Some power from these areas is transmitted southwards by the National Grid. Coastal power-stations, such as the Isle of Grain in the Thames Estuary, are favoured by open areas of land for building, nearby water for cooling and sea transport for oil or coal.

Today, the less efficient coal-fired power-stations are being closed down or temporarily put out of production. Modern coal-fired stations, however, will remain important electricity producers for many years to come.

Now try Exercises 23 and 24.

◀ **Fig. 66** The coalfield location of Britain's thermal power-stations

51

Exercise 23 Electricity

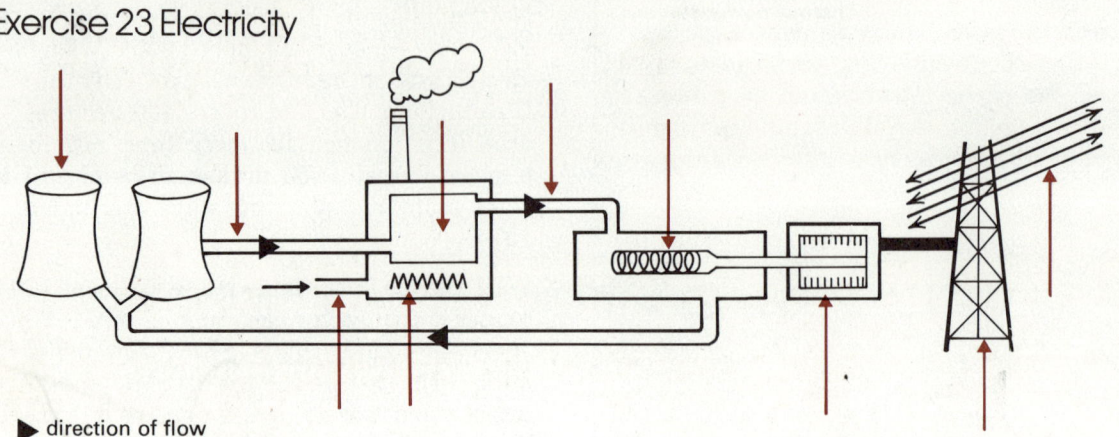

▶ direction of flow

1 Copy this simplified diagram of a coal-fired power-station and add the captions from the list below.
   cooling towers, boiler, turbine, generator, steam pipe, water pipe, burners, transmission cables, pylon, coal supply inlet.
2 Name three kinds of fuel which may be used to generate electricity.
3 Name one kind of power-station which requires no fuel.
4 What kind of movement does a turbine make:
   (a) up and down; (b) side to side; (c) upwards; (d) downwards; (e) spinning?
5 What kind of force turns the turbine in a thermal power-station?
6 Place the following in the correct order (begin with heat):
   heat, steam, generator, water, electricity, turbine, consumer, transmission.
7 What name is given to the network of electricity transmission lines?
8 Explain the difference between primary fuel and secondary power.
9 Give two reasons why generating electricity may be considered wasteful.
10 Why do many power-stations shut down for much of the day?
11 Which two of the following are not needed for building a coal-fired power-station:
   (a) large area of flat land;
   (b) water supply;
   (c) nearby coalfield;
   (d) fast-flowing streams;
   (e) a site on the outskirts of a large city?
12 Name two methods of generating electricity which use renewable sources of energy.

Exercise 24 Choose the best site

Study the map carefully. The islanders of Cola have recently discovered a coalfield which will provide long-term supplies of cheap coal. They want to build a large coal-fired power-station capable of generating most of the island's electricity needs, and they are trying to choose the best site.

1 Give two reasons why site C is unsuitable.
2 Give two advantages and one disadvantage of choosing site A.
3 Which two of the following factors help to explain why site E is unsuitable:
   (a) distance from main towns;
   (b) high cost of transporting coal by rail;
   (c) lack of cooling water;
   (d) high cost of imported oil?
4 Give four advantages of choosing site B.

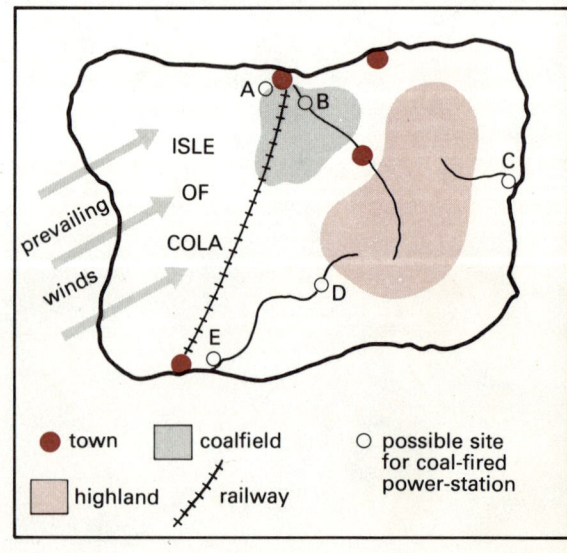

ISLE OF COLA

prevailing winds

| ● town | ▨ coalfield | ○ possible site for coal-fired power-station |
| ▨ highland | ⤬ railway | |

Hydroelectric power

Before steam power, the water-wheel was widely used to drive flour-mills (Fig. 49, page 38). It also powered the earliest textile and metal-working machines of the Industrial Revolution. *Hydroelectric power-stations* harness today's water-power.

The gigantic Hoover Dam (Fig. 67) which hólds back the waters of the River Colorado in the United States, incorporates a hydroelectric power plant. This provides power for a large area of south-western USA.

Fig. 68 shows how hydroelectricity is produced.

Fig. 67 The Hoover Dam provides electricity and irrigates land in California, Arizona and Mexico

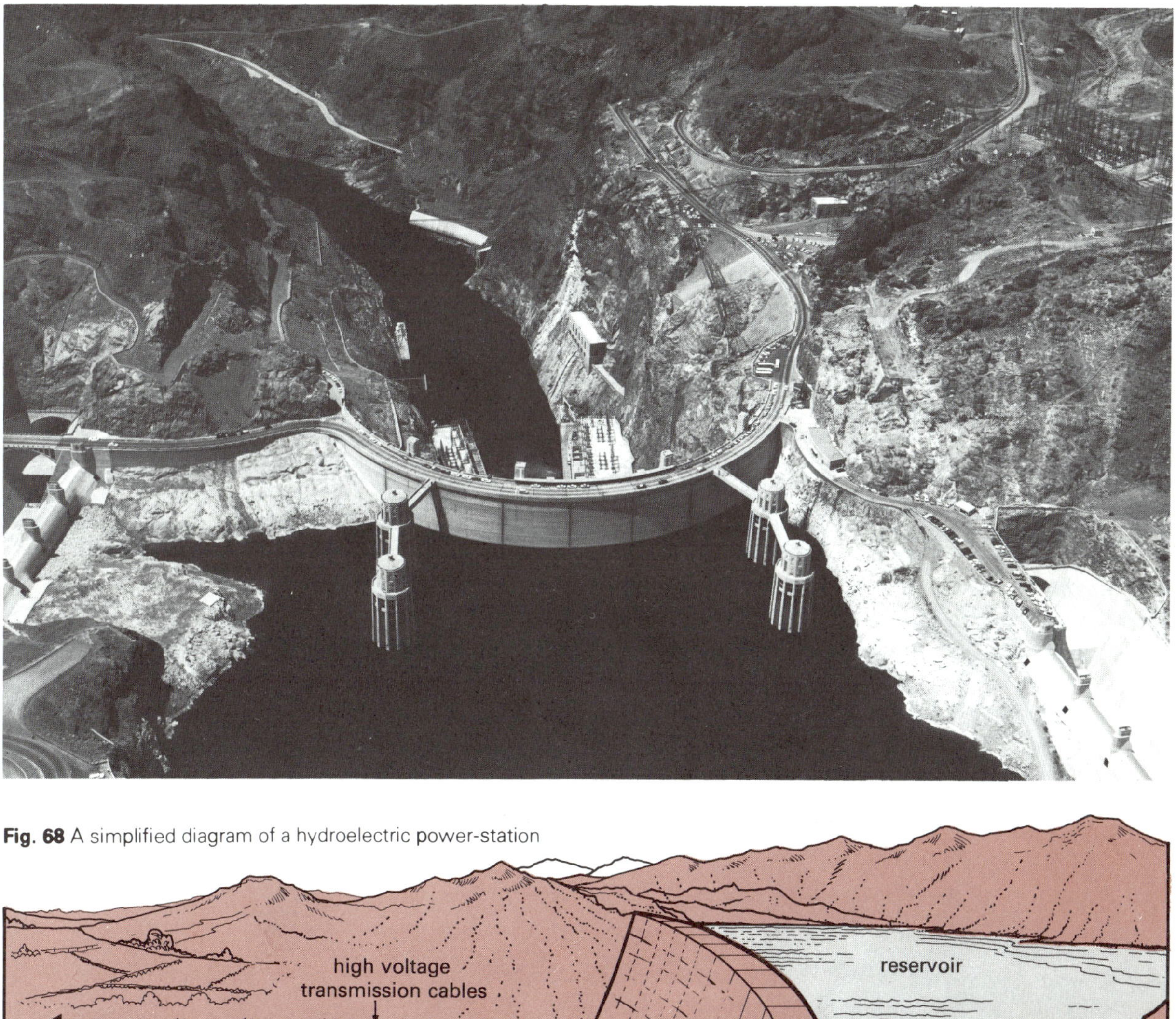

Fig. 68 A simplified diagram of a hydroelectric power-station

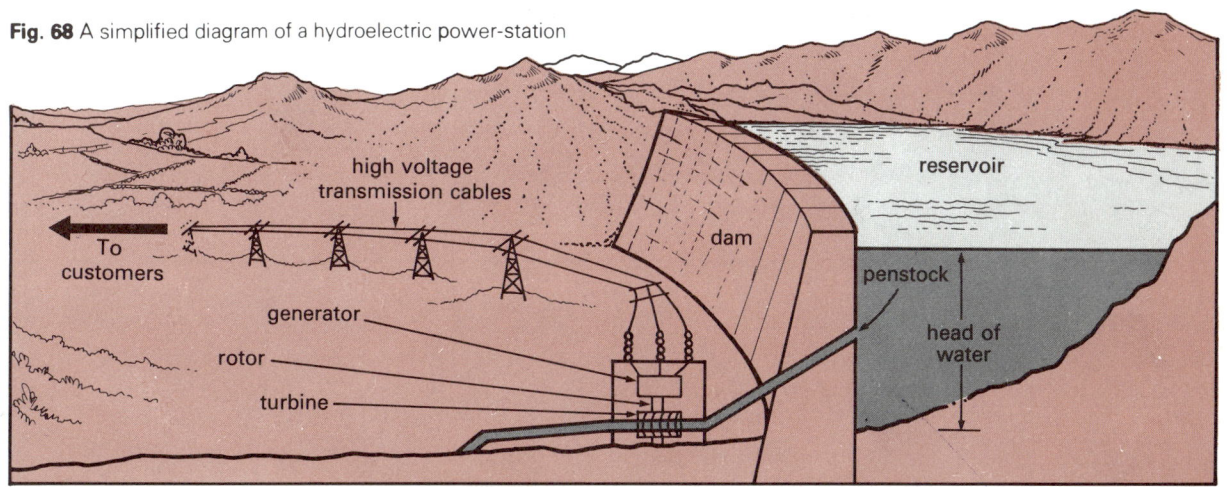

Power is generated by turbines installed at the base of the dam. Water rushes from the lake behind the dam down large pipes called *penstocks*. The force of water spins the blades of the turbine which then operates the rotor to generate electricity. The amount of power depends upon the quantity of water passing through the turbines and the vertical distance between the level of the turbines and the surface of the lake – a measurement known as the *head* of water.

Building HEP stations

The Kemano HEP scheme in British Columbia (Fig. 20, page 16) supplies the power used to smelt aluminium at Kitimat. The river Nechako, which once flowed eastwards from the coastal moun-

tains, provides water to operate the turbines. The Kenney Dam was built across the river to create a large lake. The water in the lake is diverted 16 kilometres westwards to the Kemano power-station by tunnels blasted through the coastal mountains. The power-station is built in a gigantic man-made cavern 500 metres inside a mountain (so that it is safe from bombs in case of war). The turbines are built at a level 800 metres below the lake's surface.

Schemes such as Kemano are enormously expensive. Before building begins, the area must be studied by scientists and engineers to make sure that rainfall is sufficient and reliable; and that the rocks are strong enough to take the great weight of the dam and the power-station. Power lines are often built over difficult terrain to deliver electricity to consumers. Although this means that the

Fig. 69 The Snowy Mountain HEP and irrigation scheme

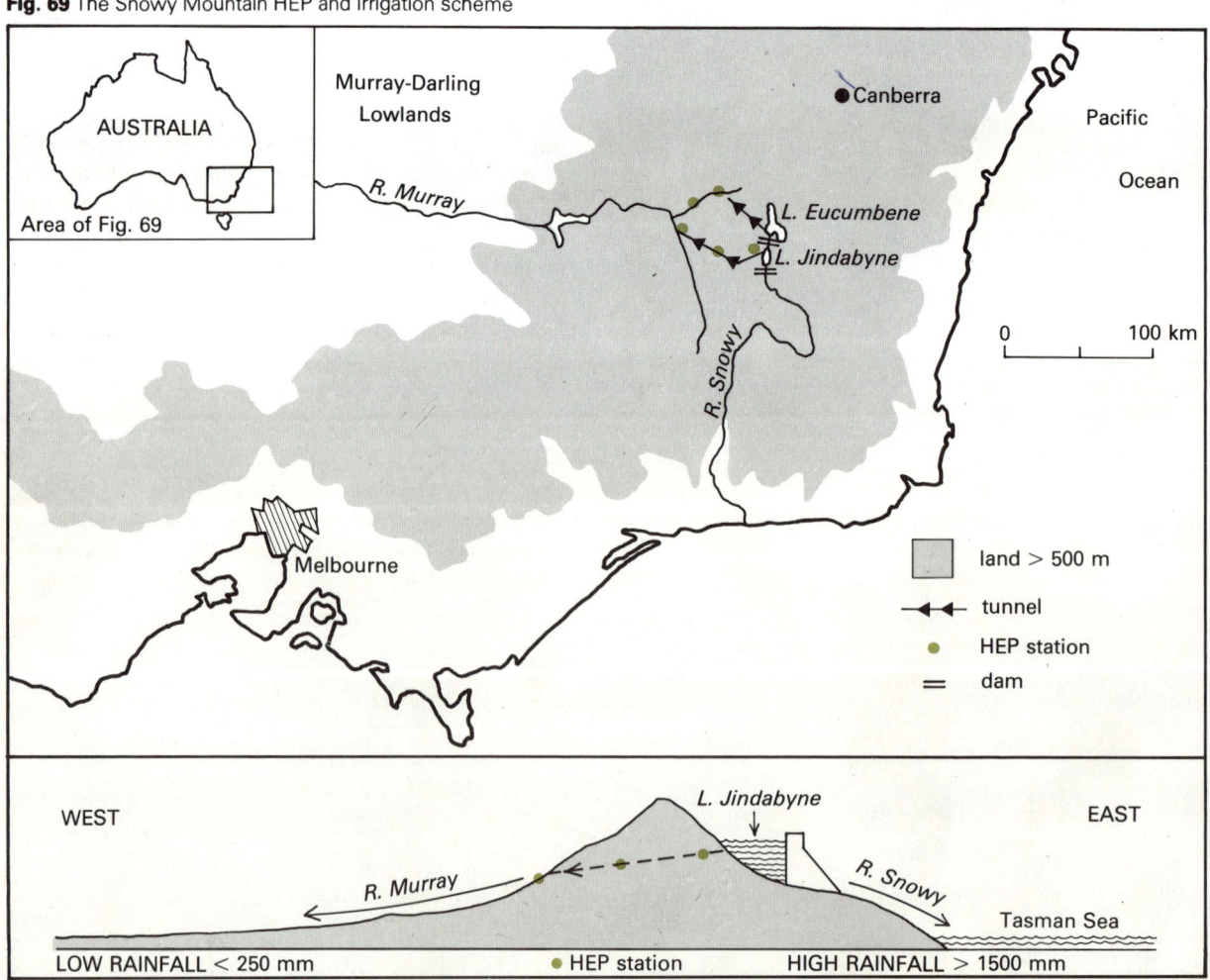

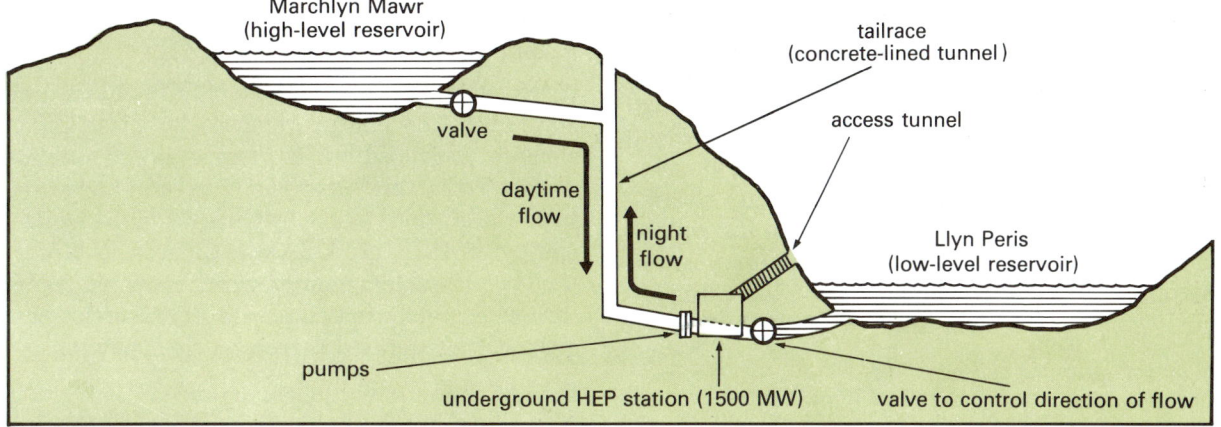

Fig. 70 A cross-section of the Dinorwic pumped storage scheme

cost of constructing HEP stations is high, running costs are lower than thermal stations because HEP needs no primary fuel except falling water.

Multi-purpose HEP

Because of the high costs of construction, many HEP schemes do not provide power alone. The Snowy Mountain Scheme in Australia provides irrigation water for over 5000 square kilometres of semi-desert in the Murray-Darling Lowlands (Fig. 69). In North America, the Tennessee Valley Authority uses dozens of dams to control flooding and soil erosion, and these also provide recreational facilities such as fishing and sailing. The power generated by the scheme has done a great deal to boost industry in what was once a poor, backward part of the country. Further north, the dams of the power project of the St. Lawrence Seaway provide valuable electricity and enable ocean-going vessels to reach the heart of the continent by connecting the North Atlantic with the Great Lakes. Power generated here is especially important since the St. Lawrence flows through the most densely populated part of North America.

Pumped storage

Hydroelectricity offers a method of 'storing' power. A *pumped storage* scheme uses electricity generated at night when demand is low to pump water to a high-level lake. During daytime periods of peak demand, this water is allowed to fall through the turbines to make power. It is then collected and stored in a low-level lake until night-time when pumps raise it once more to the high lake. The process is illustrated by Fig. 70 which is a simple diagram of the Dinorwic scheme in North Wales.

Fig. 71 The hydrological cycle

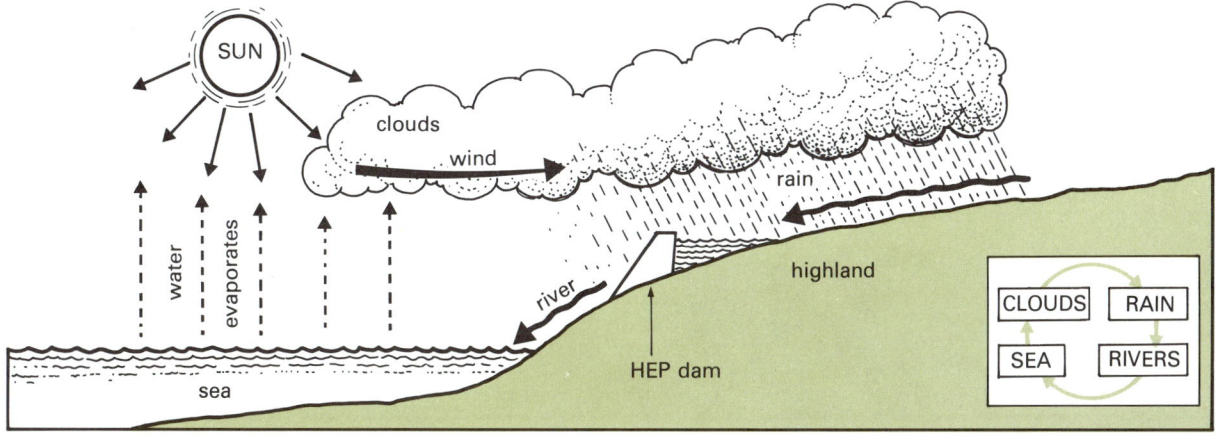

Hydroelectric power has the vital advantage of being a *renewable resource*. Coal and oil are non-renewable but HEP can be produced as long as there is sufficient water available. Its energy depends upon the *hydrological cycle* shown in Fig. 71. The sun evaporates water, the water vapour forms clouds, and the winds blow the rain clouds over the highlands. Here rain falls and the water collects to form a source of power. As fossil fuels become scarce and even more costly, the value of HEP is certain to increase.

Now try Exercises 25 and 26.

Nuclear power

In 1945, atomic bombs of awesome power destroyed the Japanese cities of Hiroshima and Nagasaki. The energy released by these explosions was caused by a process called *nuclear fission*. Today, nuclear fission has been harnessed in power-stations which produce 6% of the world's energy. In 1978, the USA was the world's largest producer of nuclear energy, which provided 12% of the nation's electricity. In Britain, 13% of the total electricity generated is from nuclear power.

Fig. 72 shows how nuclear power-stations differ from conventional plants. Instead of burning coal, oil or natural gas, the heat from nuclear fission is used to boil water and generate steam. Nuclear fission is a process in which atoms of a radioactive substance such as *uranium* or *plutonium* are split

---

Exercise 25 Water-power

Complete the passage using the ten correct words chosen from the list.

flour-mills, hydroelectric, machines, water-wheel, sea, sun, force, boilers, turbines, rainfall, building, operating, plentiful, renewable, fixed, scarce.

The _____ was used to produce power to drive _____ and _____ during the early part of the Industrial Revolution. Today, _____ power-stations use the _____ of moving water to drive _____. HEP stations are best situated in mountainous areas of regular and plentiful _____. Although _____ costs are low, installation costs are expensive. Because it is a _____ energy resource, HEP is likely to become more important as supplies of oil become _____.

Exercise 26 Hydroelectricity

Study the map carefully. It shows five possible sites for building a hydroelectric power-station to supply all the towns on the island.

1 Explain two advantages and one disadvantage of site A.
2 Which site is most likely to be affected by:
   (a) winter freezing; (b) highest transmission costs; (c) too little water; (d) flooding?
3 a Give three advantages of choosing site B.
   b Give one possible disadvantage of choosing site B.

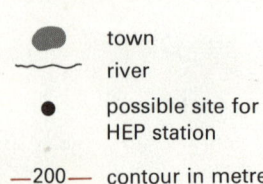

town	
river	
possible site for HEP station	
—200—	contour in metres

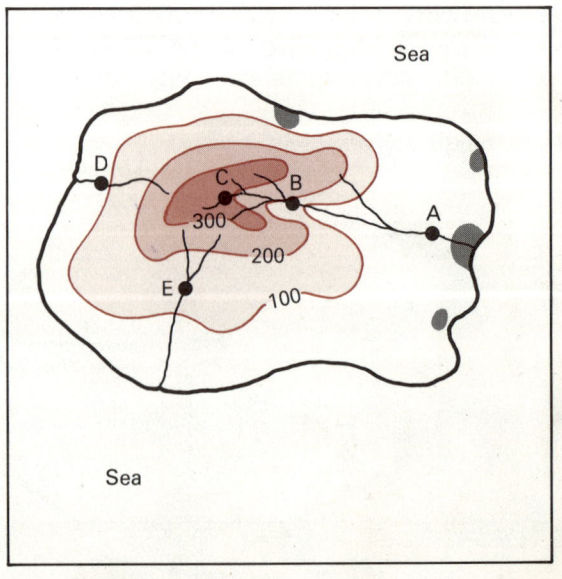

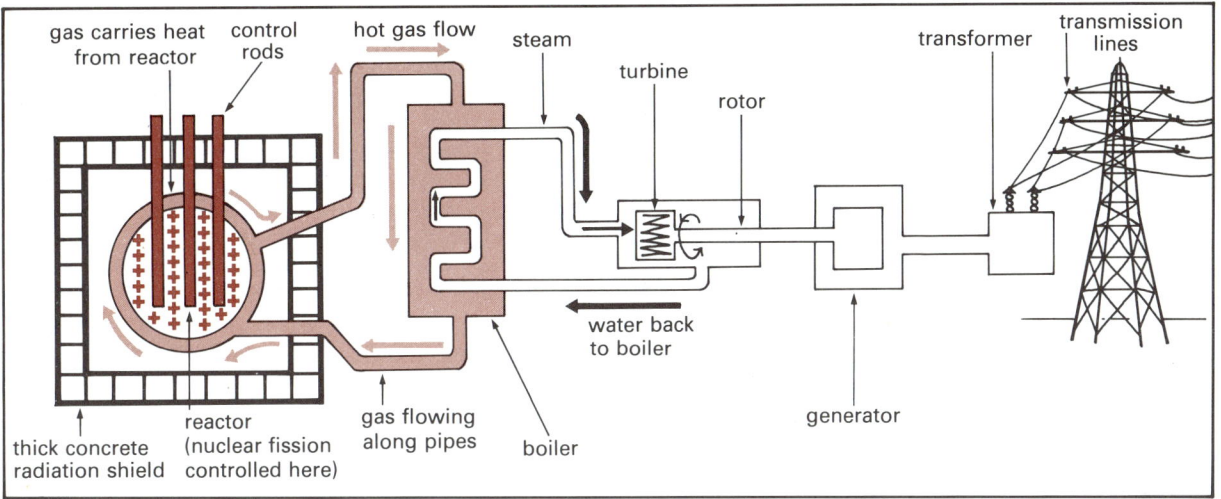

Fig. 72 A simplified diagram of a nuclear power-station

apart. As they split apart, particles called neutrons and energy in the form of heat are released. Some of the escaping neutrons collide with other atoms, splitting them and so causing a *chain reaction*. In a nuclear power-station this process takes place in a *reactor*. This consists typically of a vertical cylinder 15 metres in diameter and 10 metres high. It is made of graphite blocks and is surrounded by a stainless steel shell 10 centimetres thick. This in turn is contained inside a protective shield of concrete 3 metres thick, which is necessary to contain the highly dangerous *atomic radiation*.

As the chain reaction takes place, the energy produced by the nuclear fission heats a gas which circulates around the reactor. This very hot gas in turn is used to boil water to make the steam which drives the turbines of the generating plant. To control the chain reaction, and therefore prevent a catastrophic explosion, control rods are lowered into the reactor. These are made of a material such as boron steel which absorbs some of the escaping neutrons and so reduces the rate of fission.

Nuclear power in Britain

Fig. 73 maps the distribution of nuclear power-stations in Britain. All except one have been built on coastal sites close to supplies of sea water used for cooling purposes and where solid foundations for the enormous weight of the installation can be

guaranteed. The earliest stations such as those at Calder Hall and Dounreay were built in remote areas as a result of public concern over safety. Today, stations such as Oldbury and Berkeley on the Severn estuary have been built closer to the large towns (Bristol and Cardiff) to which they supply power. Nevertheless, extremely high levels of safety control must be observed and maintained if the hazards associated with nuclear power are to be avoided.

Fig. 73 Britain's nuclear power-stations

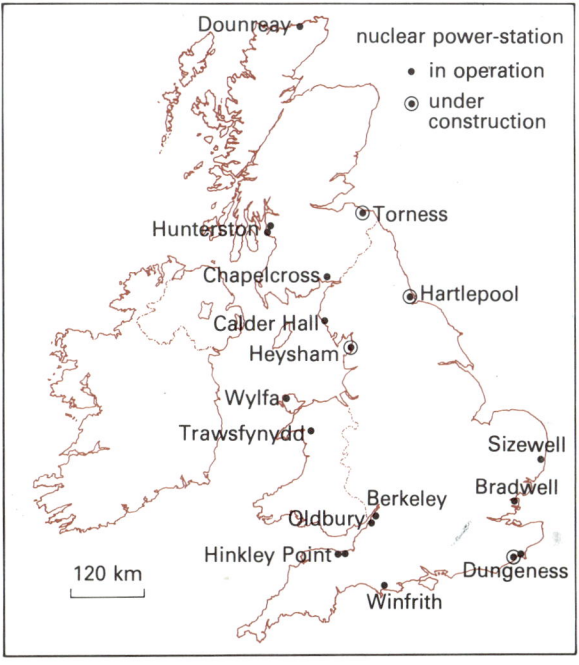

The future for nuclear power

One advantage of nuclear power is the small quantity of fuel needed. One kilogram of uranium can give as much heat as 100 000 tonnes of coal. By the year 2000, however, world nuclear power output is expected to have increased threefold. For output to increase to this extent, 0·5 million tonnes of nuclear fuel will be needed each year. The total known uranium reserves amount to about 4 million tonnes, so shortages of uranium fuel are inevitable.

Two alternative forms of nuclear energy are therefore being investigated:

1. In *nuclear fusion,* energy is released when atoms of deuterium and tritium (both plentiful) are joined together. The process produces no dangerous waste. Scientists have created energy in this way – in the form of the hydrogen bomb – but so far they have been unable to control fusion for the production of power.

2. *Breeder reactors* are being developed which are capable of producing more fuel than they consume, doubling the amount of fissionable material every ten years. Unfortunately, breeder reactors use *plutonium,* the material used to make nuclear weapons. It is even more dangerous than uranium.

There are grounds for concern about the safety of all nuclear reactors. No reactor is perfectly sealed. There is always the possibility (however slight) of an accident which might release highly toxic radiation into surrounding areas. Atomic radiation can cause cancers, deformed babies and death. The problem of the safe disposal of radioactive waste has not yet been solved. Every 1000 megawatt reactor produces about 9 cubic metres of dangerous waste each year. It takes many thousands of years to become harmless. Burial in thick concrete or glass containers out at sea or in old deep mineworkings is one method of disposal, but the associated risks are unacceptable to many people. The safety problems of breeder reactors are likely to prove the most difficult, partly because of the large amount of radioactive material created and also because plutonium is the most toxic substance known.

Although there are dreadful risks with nuclear energy, there are benefits. Nuclear power-stations do not pollute the atmosphere with smoke or damage the environment in the way coal-mining does. As supplies of coal and oil run out, we must either learn to live with less energy or find new ways of making electricity. Unless we learn to harness more fully the energy of the sun, the wind and the waves, nuclear energy is likely to provide steadily increasing amounts of power.

Now try Exercise 27.

Alternative sources of energy

As North Sea oil and gas production declines in about 30 years' time, Britain's power supplies will become increasingly dependent upon coal reserves which are expected to last for at least 250 years. Severe increases in the cost of energy since 1970 have prompted research into alternative sources of energy, particularly renewable sources such as solar, wind, tidal and wave power; and the conversion of plant tissue (biomass) into fuel.

Exercise 27 Nuclear energy

Answer true or false.

1 Nuclear fission produces the heat needed by a nuclear power-station.
2 The turbines in a nuclear power-station are driven by an atomic reactor.
3 One kilogram of uranium provides more energy than 10 million tonnes of coal.
4 Nuclear power may become a major source of energy in the twenty-first century.
5 The safe disposal of nuclear waste is a very serious problem.
6 The first nuclear power-stations were built in isolated positions for safety reasons.
7 Most nuclear power-stations in Britain have been built near the sea.
8 There are no nuclear power-stations within 100 kilometres of London.
9 A coastal location is needed so that nuclear fuel can be easily imported.
10 Electricity produced by nuclear fusion can now be supplied.
11 Plutonium is a safer nuclear fuel than uranium.
12 Supplies of fuel needed by breeder reactors will be exhausted in 10 years' time.

Fig. 74a A 1000-kW solar furnace at Font-Romeu in the Pyrenees, S W France

Fig. 74b How mirrors are used to focus the sun's rays onto the furnace ▼

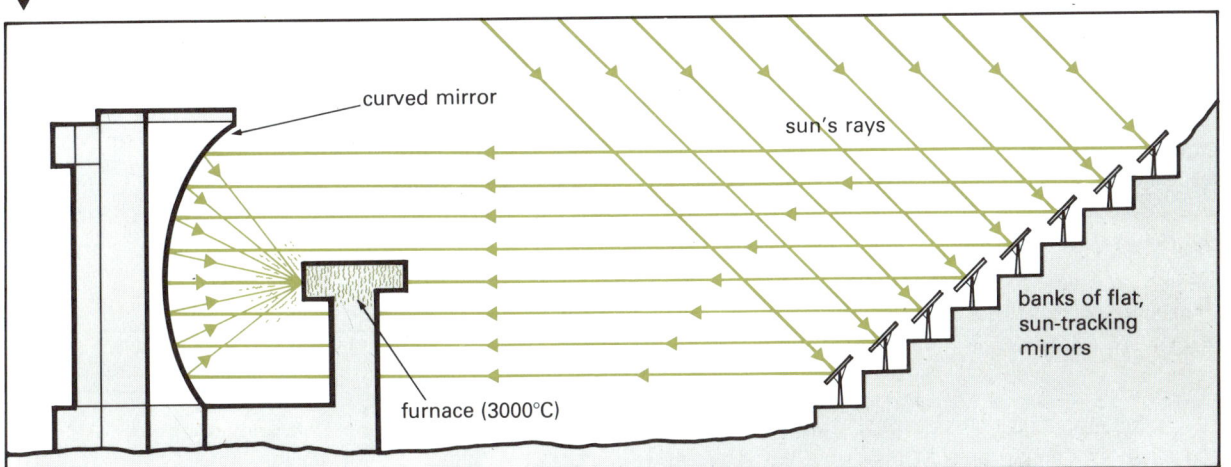

curved mirror

sun's rays

banks of flat, sun-tracking mirrors

furnace (3000°C)

Fig. 75 *Solar Challenger* powered by solar cells

Solar power (power from the sun)

Fig. 74 shows the huge curved mirrors of a solar furnace in south-west France. Intense heat can be produced by concentrating the sun's rays at a focal point. Unfortunately such costly installations depend upon long hours of continuous sunshine and their use is therefore limited.

The solar-powered aeroplane (Fig. 75) uses electricity generated by sunlight falling upon thousands of solar cells carried by the plane. Like the solar furnace it is a scientific experiment not in commercial use. Solar roof panels, however, already boost existing heating systems in some homes and workplaces.

Fig. 76a The barrage across the Rance estuary houses the world's first operational tidal power-station

Fig. 76b The location of the Rance tidal power-station

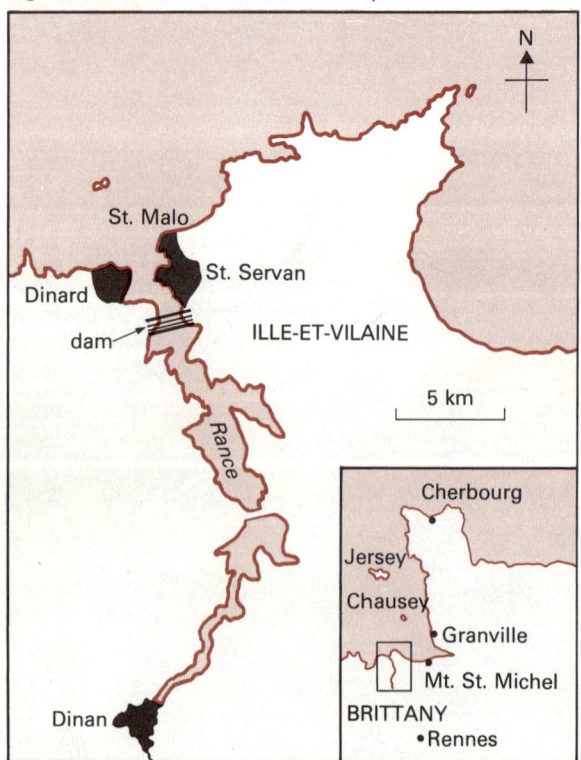

Fig. 77 How the tide turns the special turbines

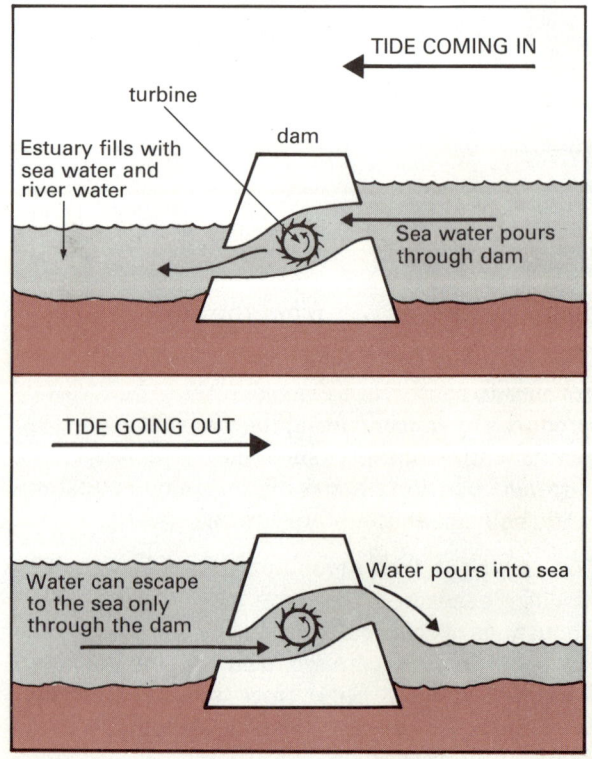

Sea power: Electricity from the tides

Every day the oceans rise and fall with the tides. Some coastal areas, especially those with funnel-shaped inlets, have a daily tidal range of up to 10 metres. The power-station at St. Malo in Brittany, France, (Fig. 76) is the world's first *tidal power-station*. Fig. 77 shows how special reversible turbines generate electricity from both incoming and outgoing tides. In addition, the dam across the Rance estuary links the towns of St. Malo and Dinard with a new road which cuts the journey from 40 kilometres to 5 kilometres. As with HEP, tides offer a cheap, continuous method of generating power. Rising fossil fuel prices are therefore encouraging interest in similar schemes proposed for the Severn estuary in Britain and for the Bay of Fundy in eastern Canada.

Wave power

The prevailing south-westerly winds blowing uninterruptedly across the North Atlantic have created some of the world's most energetic wave sites along the western approaches of the British Isles. The greatest potential energy is available in winter when it is most needed. Scientists believe that if machines could be developed to extract only a small fraction of this energy, wave power could make a useful contribution to Britain's electricity supplies. The main research devices include 'Salter ducks' and 'Cockerell rafts'. These machines convert their random rocking motions into a rotating movement capable of generating electricity. Current tests are being carried out using 1:100 scale models, and there are plans for trials in Loch Ness and the Solent using 1:10 scale models. If wave energy is to be successful, machines must be capable of withstanding the hostile conditions of the northern seas. Massive investment on a scale greater than North Sea oil and gas exploration will be necessary to launch such a project.

Geothermal power

In New Zealand and Iceland sources of underground heat are used to generate electricity and for the direct heating of homes and factories. Fig. 78 shows a geothermal power-station at Wairakei in North Island, New Zealand, where molten rocks beneath the surface convert ground water into steam. This is piped to turbine generators to

Fig. 78 Part of the geothermal borefield at Wairakei

make electricity. Geothermal steam helps to keep heating costs low in sub-arctic Iceland. Fig. 79 shows glasshouses heated in this way. In Britain, underground reservoirs of hot water at 60 to 70°C are used at Marchwood power-station near Southampton. The hot water is fed into the boilers of the thermal power-station to reduce heating costs.

Fig. 79 Geothermal steam is used to heat these glasshouses in Hverageroi, the flower-growing centre of southern Iceland

Fig. 80 A 2.5-MW wind turbine at Medicine Bow, USA

Similar reservoirs of underground hot water are used to heat blocks of flats in the Villeneuve district of Paris.

Geothermal power is unlikely, however, to supply more than a small fraction of total world energy because there are too few places where it is accessible.

Wind power

Windmills have been used for centuries but the modern version is a much more efficient machine capable of generating 4 megawatts of electricity (Fig. 80). Unfortunately, winds are unreliable and the windiest places, such as polar regions and deserts, are too far from the centres of population to make wind power a major source of energy.

Energy conservation

The saving or conservation of energy resources is now of major importance to all nations. Energy-saving can be achieved by everybody: buildings can be insulated with fibre glass blankets in the roof; windows can be double-glazed; small cars can be designed with low petrol consumption. In Brazil, motorists use *gasohol* instead of petrol. Gasohol is produced by fermenting sugar cane – so they can grow their fuel! The twenty-first century, however, is likely to see coal and nuclear power supplying an increasing proportion of the world's energy as oil becomes even more costly.

Now try Exercise 28.

Exercise 28
Forms of power

1 Match each of the following forms of energy with the correct description:
   (a) thermal   uses mirrors
   (b) solar   burns coal or oil
   (c) HEP   is always sited on the coast
   (d) nuclear   depends on rainfall
   (e) tidal   uses underground heat
   (f) geothermal needs a reactor

2 Name the 'odd one out' in each list.
   (a) paraffin, diesel oil, uranium, petrol, oil.
   (b) Sizewell, Oldbury, Dounreay, Calder Hall, West Burton.
   (c) coal, oil, natural gas, uranium, HEP.
   (d) timber, oil, peat, coal, lignite.

3 From the list of energy sources, choose the correct one for each of the following:
   (a) internal combustion engine; (b) jet engine; (c) nuclear power-station; (d) HEP; (e) steam-engine; (f) solar cell; (g) tidal power-station; (h) sailing ship.
   Energy sources
   paraffin, petrol, sea, wind, water, uranium, coal, sun.

4 Which two of the following factors make wave power a potential source of energy for Britain?
   (a) Wave power is free.
   (b) Around Britain, the greatest wave energy is in winter.
   (c) Wave power is a renewable resource.
   (d) South-westerly winds blow daily across the North Atlantic.

5 Which form of energy was first used as a weapon?

6 Name four kinds of fuel extracted from the ground.

7 Name four kinds of fuel which may be burned in a thermal power-station.

8 Name four kinds of power which cause no air pollution.

9 What is OPEC? Give two reasons to explain its importance.

10 Suggest two ways in which energy can be saved (a) at home; (b) in transport.

11 Which two of the following sources of power are likely to increase in importance by the year 2000:
   (a) oil; (b) HEP; (c) natural gas; (d) nuclear energy?

12 Which three forms of primary energy together supply over 90% of Britain's needs?

Chapter 3 Going Places

Personal journeys

Every day, people make journeys. In the home, we make short but frequent journeys to different rooms used for various activities such as preparing food, eating, sleeping, bathing or watching television. Each journey is made in order to satisfy a particular need. Longer journeys are necessary to satisfy our need to buy goods (shopping), for education (school), to obtain medical care (hospital or doctor), for recreation (swimming-baths, playing-fields, etc.).

Fig. 81 is a diagram of journeys made during the course of a week by the King family. Each journey is shown as a straight line which links the family home directly to the destination. This line represents the shortest possible path between the origin and destination of each journey; it is known as a *desire line*. The actual route taken is less direct in most cases. Note how each journey is made in order to satisfy a personal or family need – Mrs King works at the Health Centre, Mr King in an office, the children go to school.

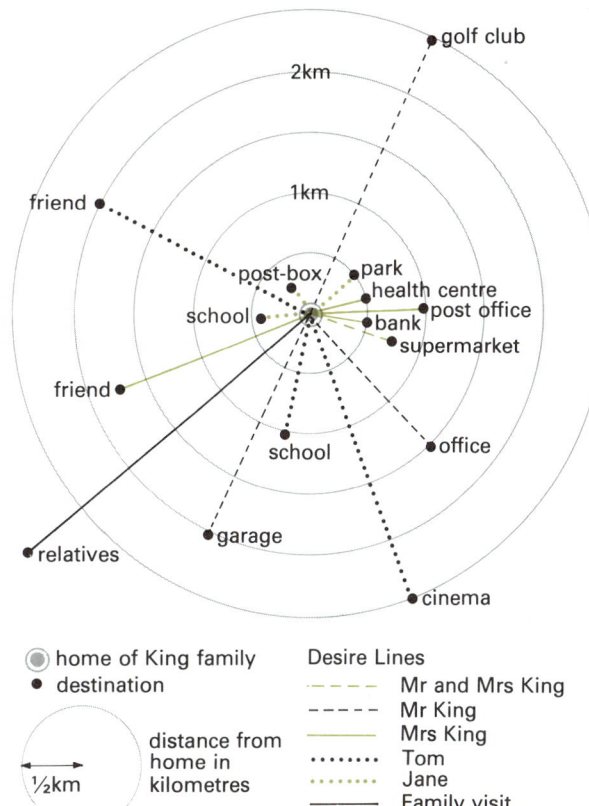

Fig. 81 The personal journeys made by the King family

Subject	Purpose	Destination	Desire line distance (shortest path)	Mode	Frequency
Mr King	work	office	1.5 km	car	each weekday
Mr King	car repair	garage	2 km	car	twice yearly
Mr King	recreation	golf club	2.5 km	car	twice weekly
Mr and Mrs King	shopping	supermarket	0.7 km	car	once weekly
Mrs King	buy stamps	post office	0.95 km	bus	once monthly
Mrs King	social	friend's house	1.7 km	car	once weekly
Mrs King	work	health centre	0.5 km	walk	each weekday
Mrs King	banking	bank	0.5 km	walk	once weekly
Tom	education	school	1 km	bicycle	each weekday
Tom	social	friend's house	2 km	bicycle	twice weekly
Tom	entertainment	cinema	2.5 km	bus	once monthly
Jane	education	school	0.4 km	walk	each weekday
Jane	post letter	postbox	0.25 km	walk	once monthly
Jane	recreation	park	0.5 km	walk	once weekly
Family	social	relatives	3 km	car	once monthly

The journeys differ in several important ways according to their:

1. Purpose – why the journey is made.
2. Destination – where the journey ends, and the need satisfied.
3. Route – the path taken between origin and destination.
4. The mode of travel – walk, cycle, car, bus, etc.
5. Frequency – how often such a journey is made.
6. Length – distance travelled (usually greater than desire line distance).

Route detours

Fig. 82 shows four examples of the actual *routes* taken by members of the King family. It is drawn to scale so that journeys can be measured accurately. Note that Jane's journey to school is the shortest possible path since both her home and school are situated on the same straight road. The King's route to the supermarket is less direct because they must make a *detour*. Their route *deviates* from the shortest path. Most personal journeys involve detours, so journeys are longer and more costly in terms of time and money. Journey detours can be compared by giving them an index. The index is calculated by using the following formula:

$$\text{Detour Index} = \frac{\text{Real journey distance}}{\text{Direct (shortest path) distance}} \times \frac{100}{1}$$

Example

$$\text{Detour Index of King's journey to the supermarket} = \frac{840 \text{ m}}{700 \text{ m}} \times \frac{100}{1} = 120$$

Because Jane's journey to school is direct, the detour index is 100. A detour index of 200 means that the journey is twice the distance of the shortest path. *(Continued on page 67.)*

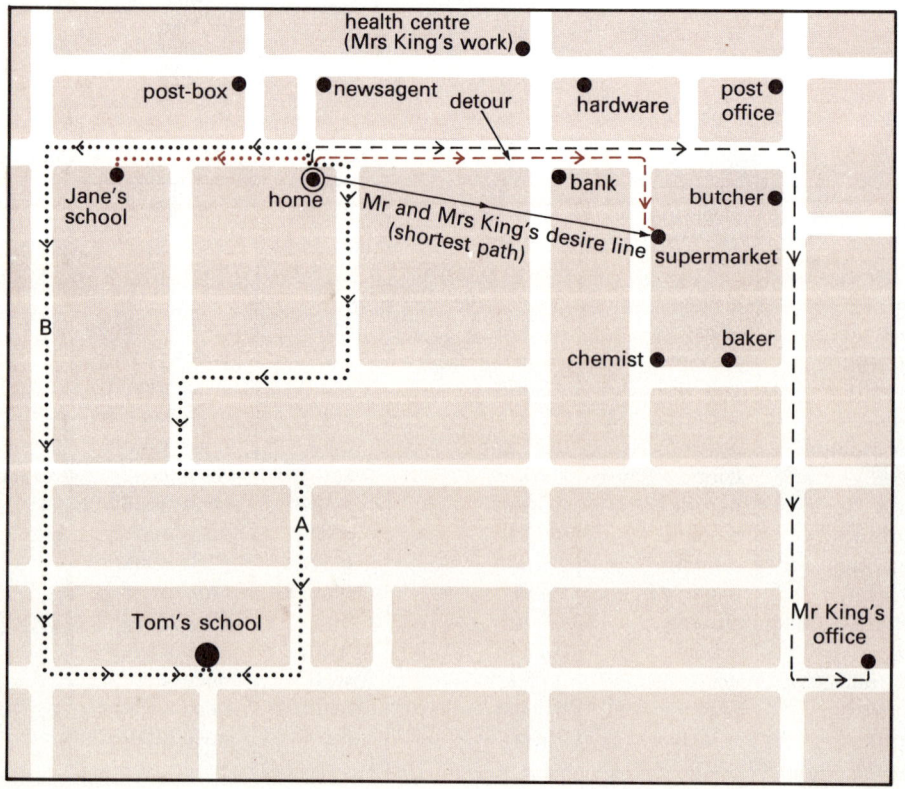

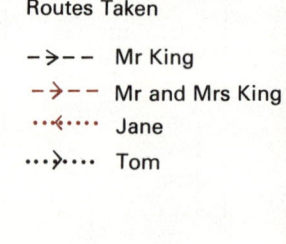

Fig. 82 Four routes taken by members of the King family

Routes Taken

->-- Mr King

->-- Mr and Mrs King

···<····· Jane

···>···· Tom

Exercise 29 The Severn Bridge

The Severn Bridge was opened in 1966 to provide a link between Chepstow and Bristol and thereby greatly improve communications in the area. This simplified map shows the road network which serves the Severn estuary region.

1 Copy and complete the table to show the distances between the selected towns in the area. The figures given are the distances via Gloucester before the bridge was built, and the distances via the bridge.

	Bristol		Bath		Swindon		Taunton	
	Via G'ster	*Via* Bridge	*Via* G'ster	*Via* Bridge	*Via* G'ster	*Via* Bridge	*Via* G'ster	*Via* Bridge
Hereford	106	79						
Chepstow	101	21						
Newport								
Cardiff								
Merthyr								

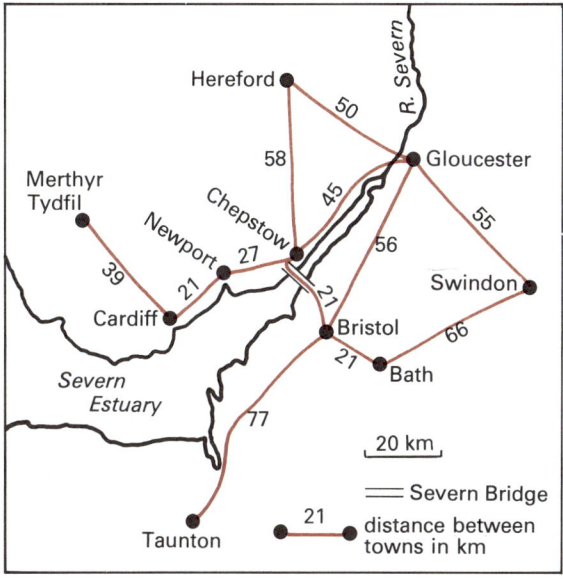

Hereford
50
58
R. Severn
Gloucester
45
55
Merthyr Tydfil
Chepstow
Newport
56
27
39
21
21
Swindon
66
Cardiff
Bristol
Severn Estuary
21
Bath
77
20 km
Severn Bridge
21
distance between towns in km
Taunton

2 Name the towns on the shortest route from:
(a) Gloucester to Bath; (b) Cardiff to Swindon; (c) Hereford to Bath; (d) Hereford to Swindon; (e) Bath to Cardiff.

3 Study your distance table carefully. Name two journeys that could be made via the Severn Bridge, but which would be shorter via another route.

4 **a** Calculate the total distance saved from Taunton to Cardiff by using the Bridge rather than the old route via Gloucester.
b Calculate the saving in time and petrol if a vehicle has an average speed of 80 km per hour and does 10 km to the litre.

5 Calculate the detour index of the journey from Taunton to Cardiff (a) via Gloucester, (b) via the Severn Bridge.

Exercise 30 Going shopping

A shopper wishes to visit the four places marked on the map. The journey must start and finish at the bus station (A). The places can be visited in any order. Letters A, B, C, D, E, F are used to identify road junctions used along the route. Distances between junctions are shown in metres.

1 Which of the following routes is the shortest?
(a) A → D → dress shop → E → department store → shoe shop → building society → C → B → A.
(b) A → B → C → building society → shoe shop → department store → E → dress shop → C → B → A.
(c) A → B → F → shoe shop → building society → E → department store → E → dress shop → C → B → A.

2 Describe the shortest route for the shopper if the visit to the shoe shop is omitted.

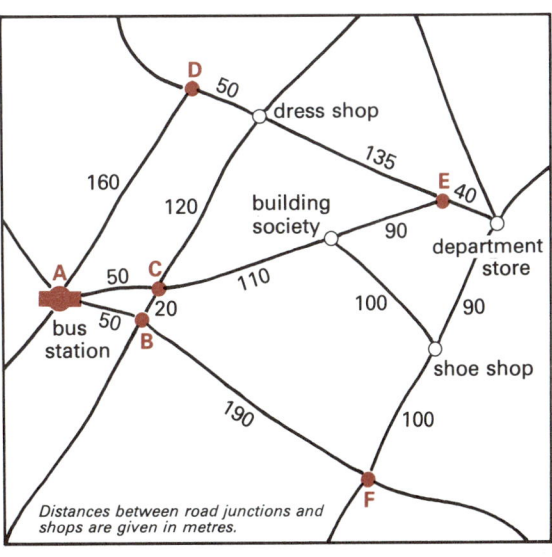

D 50
dress shop
135
160
120
building society
E 40
90
department store
A 50 C
110
100
20
bus station 50 B
90
shoe shop
190
100
F

Distances between road junctions and shops are given in metres.

Exercise 31 Going to school

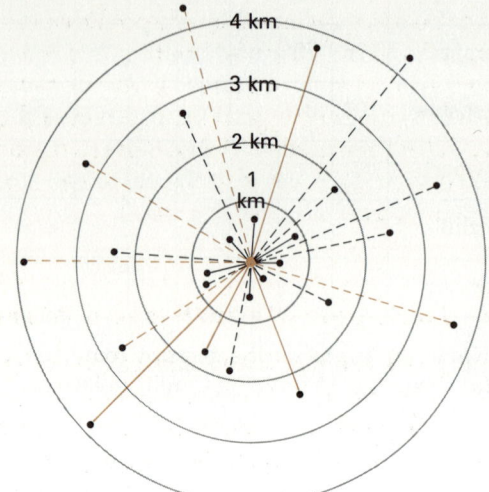

Journey distance and mode of transport

1 km
- • pupil's home   ----- cycle
- • school   —— bus
- —— walk   ----- car

	Distance of pupil's home from school				
MODE	Up to 1 km	1-2 km	2-3 km	Over 3 km	TOTAL
Walk	7				
Bicycle				2	
Bus					
Car					
TOTAL					25

The diagram shows the desire lines for journeys to school of a sample of pupils. Note how the method of transport varies according to the length of journey.

1 Copy and complete the chart. Some figures have already been included.
2 Study the diagram and your completed chart. Are the following statements about the sample of pupils true or false?
  (a) Nobody walks more than 1 kilometre.
  (b) Walking is the most common mode for those living within 2 kilometres of school.
  (c) Cycling is the most common mode of transport.
  (d) All pupils coming to school by car live over 3 kilometres from school.

3 Twenty percent of the sample go by car. What percentage
  (a) go by bus; (b) cycle; (c) walk?
4 Which of the following factors are likely to affect the type of transport used by a pupil:
  (a) school rules;
  (b) weather conditions;
  (c) distance of home from school;
  (d) month of the year;
  (e) school on parent's way to work;
  (f) family income;
  (g) price of petrol;
  (h) school bus routes;
  (i) parents' attitude to road safety;
  (j) size and weight of items to be carried?

Multi-purpose journeys

In Fig. 82, each journey had a single purpose and destination. Many personal journeys are planned so that more than one purpose can be achieved. Mr and Mrs King usually visit other places besides the supermarket when they go shopping. Note how their movements in Fig. 83 form a *journey circuit*. Such a route helps to keep the total distance as short as possible.

Fig. 83 Mr and Mrs King's journey circuit

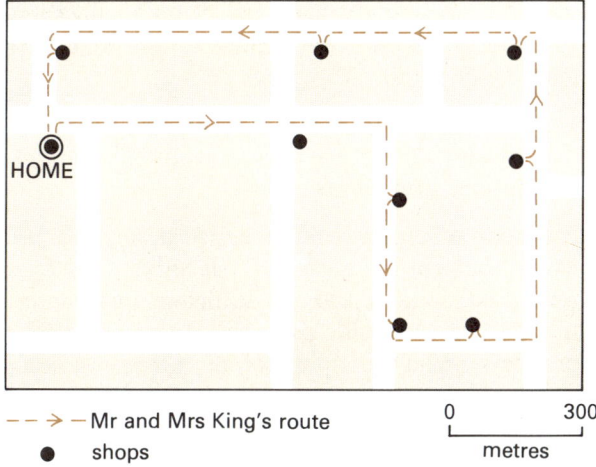

– – → – Mr and Mrs King's route
● shops

0   300

metres

Now try Exercises 29, 30 and 31.

The transport of goods

People travel journeys in order to satisfy their needs by, for example, going to work or school, and by shopping. The goods we buy must also travel from the places where they are produced (sources of *supply*) to the places where they are consumed (sources of *demand* or markets). At one time, people had to live within walking distance of the food and water they needed. Today, modern methods of transport and storage give the customer access to goods produced all over the world.

The distribution of manufactured goods

Most goods are purchased and collected by consumers from shops. Customers usually make only short journeys for low-order goods such as groceries and newspapers, but longer journeys are needed for high-order products such as jewellery, carpets or furniture. Carrying goods home, however, is only the final link in a much longer distribution chain. Fig. 84 shows the distribution chain necessary to provide the average British family with their daily bread. The journey may be divided into several stages which link the wheat farms of the Canadian Prairies with the consumer in Britain:

1. The harvested wheat is collected from the farms and taken to the local bulk storage silos.
2. The wheat is transported in bulk by rail to a port.

Fig. 84 From wheat farm to baker's shop

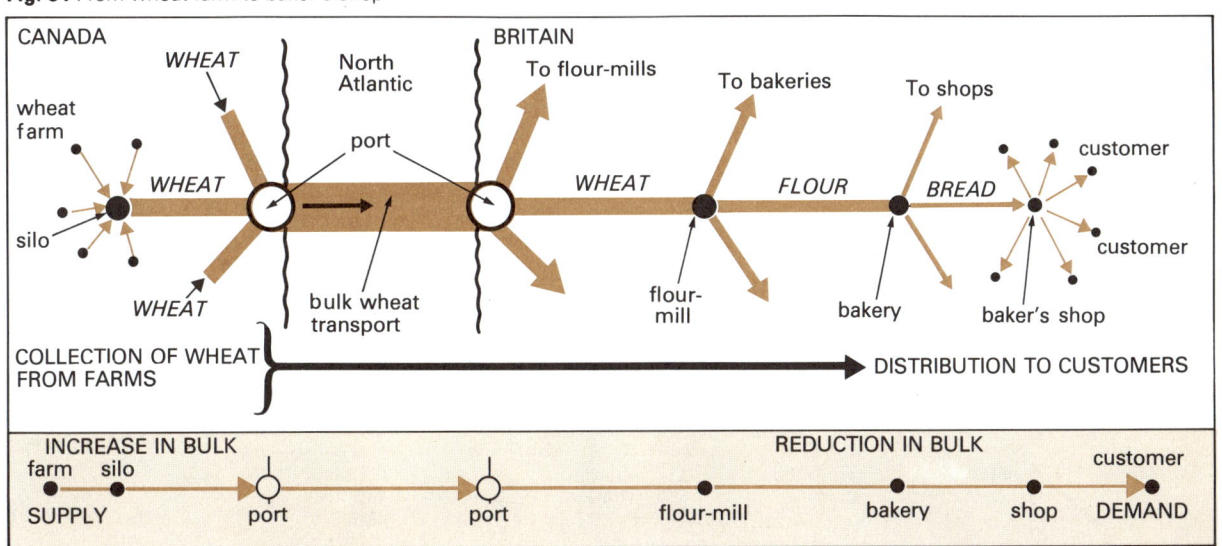

3. The grain is shipped in bulk-carriers to the British port.

4. The shipment is split into smaller loads to be carried to the flour-mills.

5. The bulk is further reduced when the flour is transported to the bakeries.

6. And the bulk is even further reduced when the loaves of bread are transported to the shops.

7. The final bulk reduction occurs when the individual loaf is carried home by the consumer.

Note how the journey is interrupted at several points:

(a) To change the mode of transport, for example, from land to sea at a port.

(b) For processing, for example, the milling of the wheat.

(c) To increase or reduce the size of the load, for example, from bulk road transport of bread to the shops to individual loaves taken home by the customer.

Water supply

Water, gas and electricity are directly conveyed to most homes through a network of pipes or cables. Personal journeys to obtain these services are not required. The supply is brought to the actual point of consumption. Such products are called *public*

utilities. Fig. 85 is a simplified diagram of the distribution of public water supplies to a local community. Note that the method of transport, the pipeline, links the reservoir (*supply*) directly to the customers (*demand*). Pipelines are very expensive to install but operating costs are low. They provide the most efficient means of transporting liquids and gases over regular routes.

The journey made by the water supply may be compared with that of Canadian wheat in Fig. 84. Both begin as natural products which are collected, processed, transported in bulk and then, step by step, broken down into smaller and more manageable units for the consumer.

Transport and trade

Prehistoric settlements depended upon food and raw materials gathered locally. Villages remained isolated and self-sufficient until better methods of transport gave people access to the products, and customers, of other regions. Improved transport enabled communities to exchange their different commodities with one another. We call this exchange of commodities *trade*. As trade developed, so living standards improved.

Fig. 85 The distribution of the public water supply

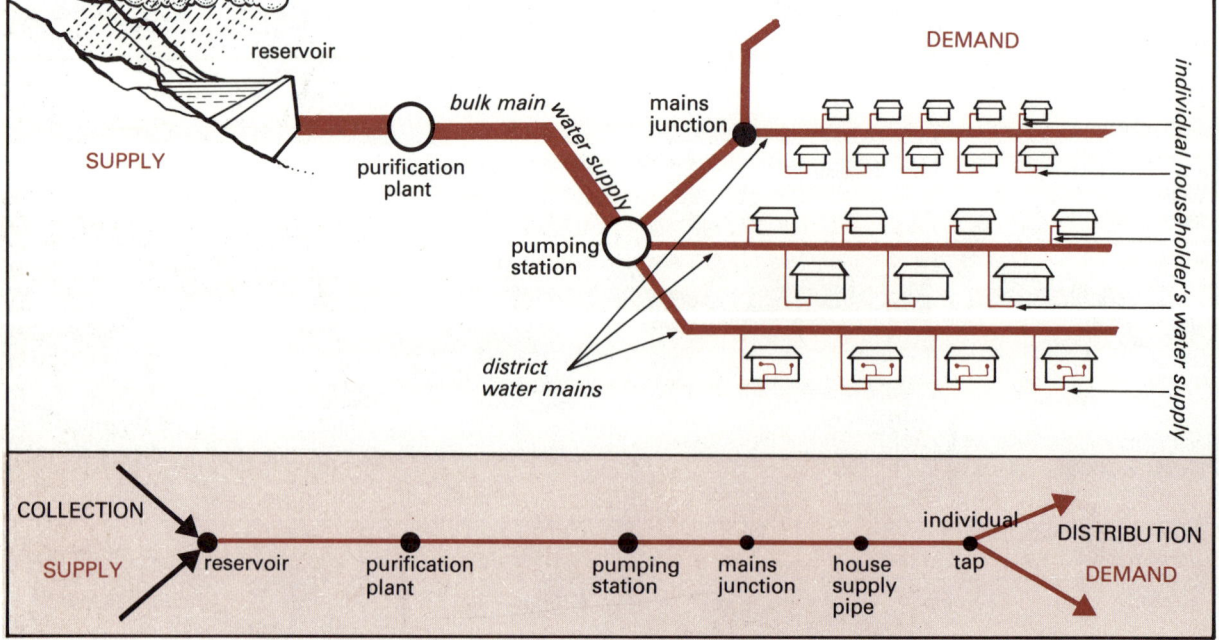

Since 1750, improvements in transport such as those seen in Fig. 86 have made possible a world-wide pattern of trade. Nations have learned to specialise in goods which they are best able to

Fig. 86
The dramatic
advances made in
transport
are illustrated by these
photographs:
a An Arab dhow, an ancient
craft, still used in the Indian
Ocean and Red Sea.
Average length – 30m;
weight – 20 tonnes
b A modern tanker
discharging its cargo at
Angle Bay, Pembrokeshire.
The world's largest tankers
are over ½ km in length
and weigh over 500 000
tonnes dwt.
c The Little Eaton tramway
in Derbyshire – an early
form of rail transport.
Average speed – 3 km/h.
d The Japanese
Shinkansen 'Bullet' train
capable of speeds of over
200 km/h.

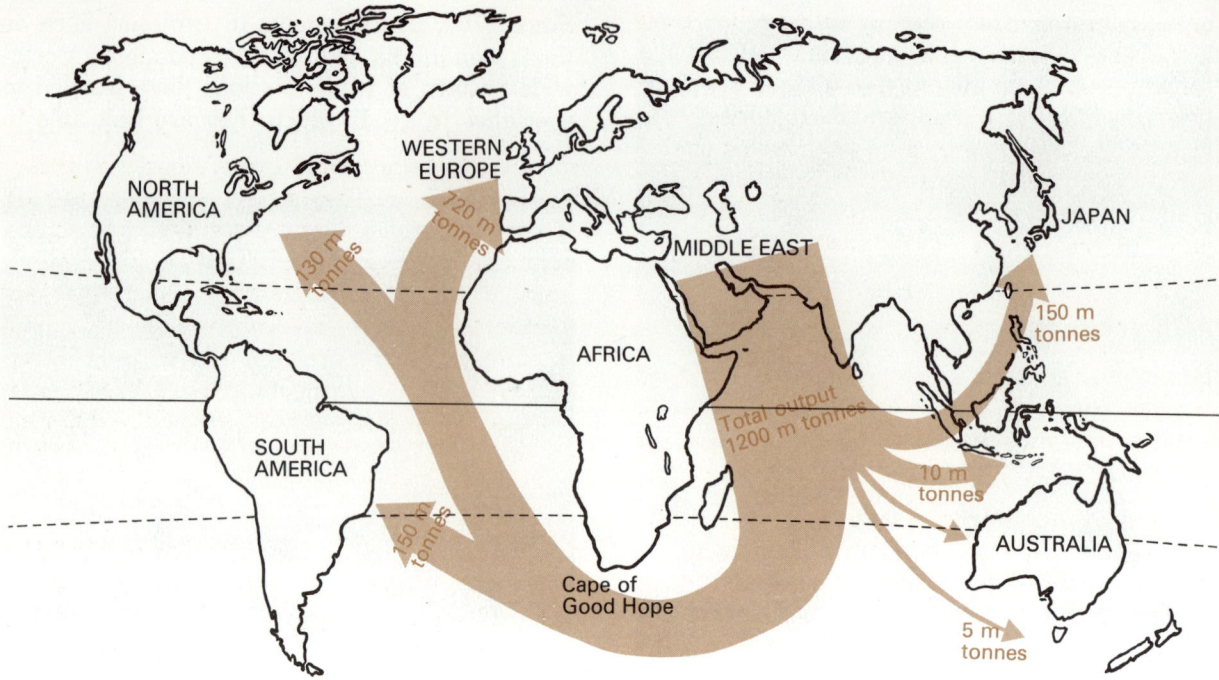

Fig. 87 World trade in Middle East oil

produce; part of their output can then be exchanged for the products of other nations. Differences in relief, climate and soils strongly influence what a nation can best produce. The dry plains of the Canadian Prairies favour profitable wheat growing; whereas tropical regions are able to grow cocoa, rubber or bananas.

Natural resources such as deposits of coal, iron-ore and oil are distributed unevenly among the continents of the world. Fig. 87 shows the world pattern of trade in oil from the Middle East. The dependence of European countries upon Middle Eastern supplies is clearly seen. Nearly half the world's shipping (by tonnage carried) is engaged in transporting oil around the Cape of Good Hope.

Today, few nations are self-sufficient in all their needs. For more than a century, Britain has been unable to feed its people from home production. To pay for imports of food and other vital raw materials such as iron-ore, cotton and timber, Britain specialises in manufacturing goods for export, competing with other industrial nations such as Germany and Japan. This vital trade between nations is dependent upon efficient world-wide transport.

Urban transport

Transport in towns and cities is essential to take people to work and to the shops. Supplies of food, water, power and consumer goods must be

Fig. 88 Morning rush-hour traffic in east London

brought in daily from regions outside the urban area. A breakdown of London's road network and railway system, for example, would quickly paralyse the capital. Millions would be unable to get to work from their homes built at increasing distances from the city centre. Closure of the Underground alone can produce chaos as extra traffic jams the roads.

A choice of transport

In Britain, a choice of transport is available for most journeys. A traveller from Manchester to London can go by private car, taxi, bus, train or plane.

The most suitable form of transport for a particular journey depends upon four major factors.

1. *Distance.* Walking is the most common way for children going to junior school and for shoppers to reach local shops. For short overland journeys, the motor vehicle is most commonly used. For long distances, a train or plane may be taken.

2. *Nature of load.* Heavy, bulky goods such as coal are transported by rail, canal or sea. Lightweight and valuable items such as jewellery and drugs are sent by road or air. Regular supplies of fluids such as oil and water are best sent by pipeline. Finished products are almost always delivered to shops by road.

3. *Urgency.* Speed is sometimes vital. Goods such as daily newspapers or perishable strawberries must be sent by overnight road or express rail transport. Urgent medical items, such as blood for transfusions or organs for transplant are sent by air and police car. Business executives, politicians and superstars, whose time is precious, fly in private aircraft to conferences, concert halls and sports events.

4. *Costs.* The average traveller usually chooses the cheapest method of transport. The cost of travel, however, cannot just be measured in terms of fuel and vehicle costs. The taxes road-users pay so that roads can be maintained must also be considered. The *social costs* incurred by the community as a result of road transport, for example, cannot be ignored. Six thousand deaths and 79 000 serious injuries occurred on the roads of Britain in 1980. Noise, air pollution by poisonous exhaust fumes, and traffic congestion (Fig. 88) are problems which seriously reduce the quality of life for many people.

Jobs in transport

Transport provides more than 10% of jobs in Britain. Fig. 89 shows the various kinds of employment connected with transport. Makers of cars, lorries, ships, trains and aircraft together employ 750 000 workers. They supply an important part of British exports. Many more have jobs in the business of moving goods and people around; nearly 1 million in road transport alone. More jobs are created in the design, construction and repair of roads, railways, docks and airports. To keep vehicles moving, more workers are needed to supply the fuel and power required by the different means of transport.

Fig. 89 Employment in Britain's transport and related industries

	Thousands
Road passenger transport	209
Road haulage contracting	215
Railways	206
Sea transport	79
Port and inland water transport	63
Air transport	89
Own account road haulage	450
Taxis	25
Crown vehicle drivers	110
Motor vehicle manufacture	424
Motor cycle and bicycle manufacture	12
Aerospace manufacture and repair	198
Railway vehicle manufacture	44
Motor repairers, distributors, filling stations, etc.	476
Total	**2600**

Finally, in an age of increasing leisure, transport is necessary for recreation and holidays near and far. Various kinds of vehicle have themselves become leisure pursuits: canoeing, sailing, motor racing, gliding, cycling and speedway are just a few examples.

Now try Exercises 32 and 33.

Exercise 32 Transport

A Choose the most suitable means of transport for the following:
1 letters from New York to London;
2 crude oil from Kuwait to Milford Haven;
3 paraffin from Fawley refinery to London Airport;
4 passenger from Waterloo Station to Liverpool Street Station;
5 coal from mine near Barnsley to Trent Valley power-station;
6 petrol from Fawley refinery to Hampshire petrol station;
7 newspapers from London to Birmingham, Preston and Glasgow;
8 letters from sorting office to homes;
9 newspapers delivered from newsagent to homes;
10 bottles of milk delivered from dairy to homes.

B From the journeys above, name five that have to be made promptly.

Exercise 33 A business trip

A business woman must travel from Nottingham to London, a distance of 200 kilometres. She must spend at least 4 hours in London to complete an important contract. She cannot leave until 09.00 and must be back in Nottingham by 18.00. She considers going by bus or by train, or driving by car. The diagram shows the journey by bus travelling at an average speed of 50 kilometres per hour. She would arrive in London at 13.00 and would be able to stay for one hour only before having to take the return bus at 14.00 to get back to Nottingham by 18.00. Bus travel is therefore not the most suitable.

Mode	Average Speed	Time *hours*	
	km/h	Travelling	in London
bus	50	8	1
car	80		
train	100		

1 Copy and complete the table. Parts of the journey by train and by car have been given.
2 At what time will she arrive in London by (a) car; (b) train?
3 At what time must she depart from London by (a) car; (b) train?
4 Choose from the following the factors most likely to favour the choice of rail travel rather than self-drive car: (a) financial cost; (b) time available in London; (c) speed; (d) safety; (e) motorway repairs on M1; (f) the need to keep to timetables; (g) parking; (h) increases in petrol prices; (i) the need to read business papers; (j) the need to arrive fresh for business.
5 Assume the executive has a private aeroplane capable of flying direct from Nottingham to the meeting in London at 400 km per hour.
   a How long could she spend in London?
   b What time would she get there if she left at 09.00?

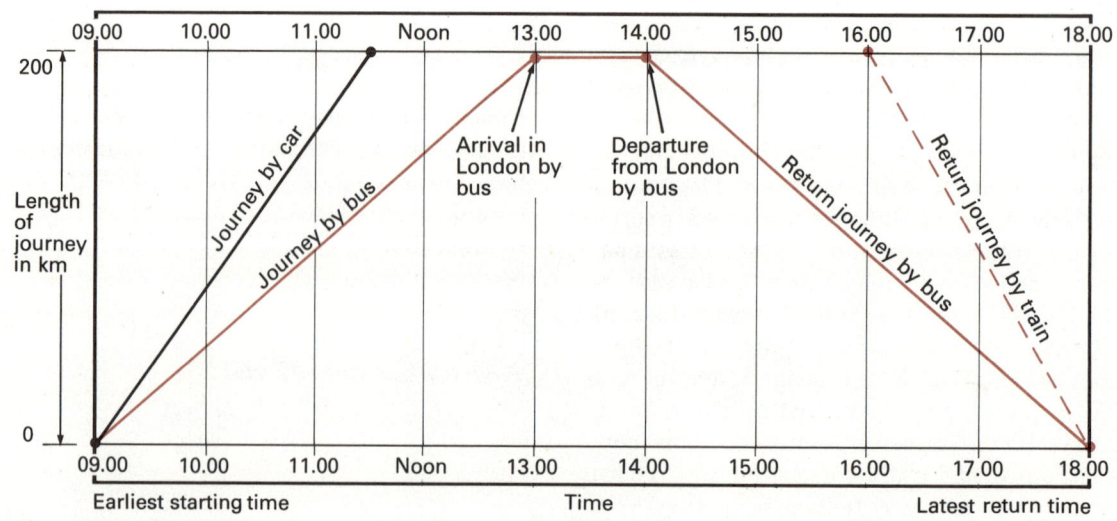

Railways in Britain

After 1920, the steady loss of rail traffic to road transport brought many British railway companies to the edge of bankruptcy. The motor vehicle offered a convenient door-to-door service with no strict timetable to observe and no fixed track to follow. Railway companies were short of money to improve services, so their businesses continued to decline. In 1947, the railways were placed under government ownership and were organised under one 'nationalised' network. To cut the huge financial losses, Dr Beeching issued his plan for 'Reshaping British Railways' in 1963. Compared with road transport, railways are faster over long distances, so Beeching decided to modernise a small number of key routes and to close or reduce services on loss-making lines.

Since the 1960s, thousands of kilometres of unprofitable lines in rural areas along holiday coasts and in declining industrial regions have been closed. Fig. 90 shows the extent of Beeching's 'axe' in the South West. Over £2000 million has been spent on railway improvements. Many of the busiest routes have been electrified, while else-where diesel locomotives have replaced steam engines (Fig. 91). Electronic signals now control trains from computerised regional centres. More durable tracks need less maintenance and allow faster, heavier and safer traffic. Express freight-liner trains carrying containerised loads (see page 94) form part of an integrated service with lorries and ships (Fig. 92). The most profitable traffic is provided by inter-city passenger services for distances over 100 kilometres. High speed, comfort and a frequent train service offer decisive advantages over cars, buses and even aeroplanes. The High Speed Train (HST) capable of 200 kilometres per hour (Fig. 93) travels from London to Bristol (180 kilometres) in 87 minutes, and from London to Newcastle (430 kilometres) in 2 hours 55 minutes. The experimental Advanced Passenger Train (APT), capable of speeds of over 250 kilometres per hour, has covered the 600 kilometres from London to Glasgow in 4 hours.

Urban railways

Urban railways provide an essential service for commuters in all the world's major cities. In Greater London, long journeys to work and serious road

Fig. 90 Rail passenger services in the South-west Peninsula in 1947 and 1981

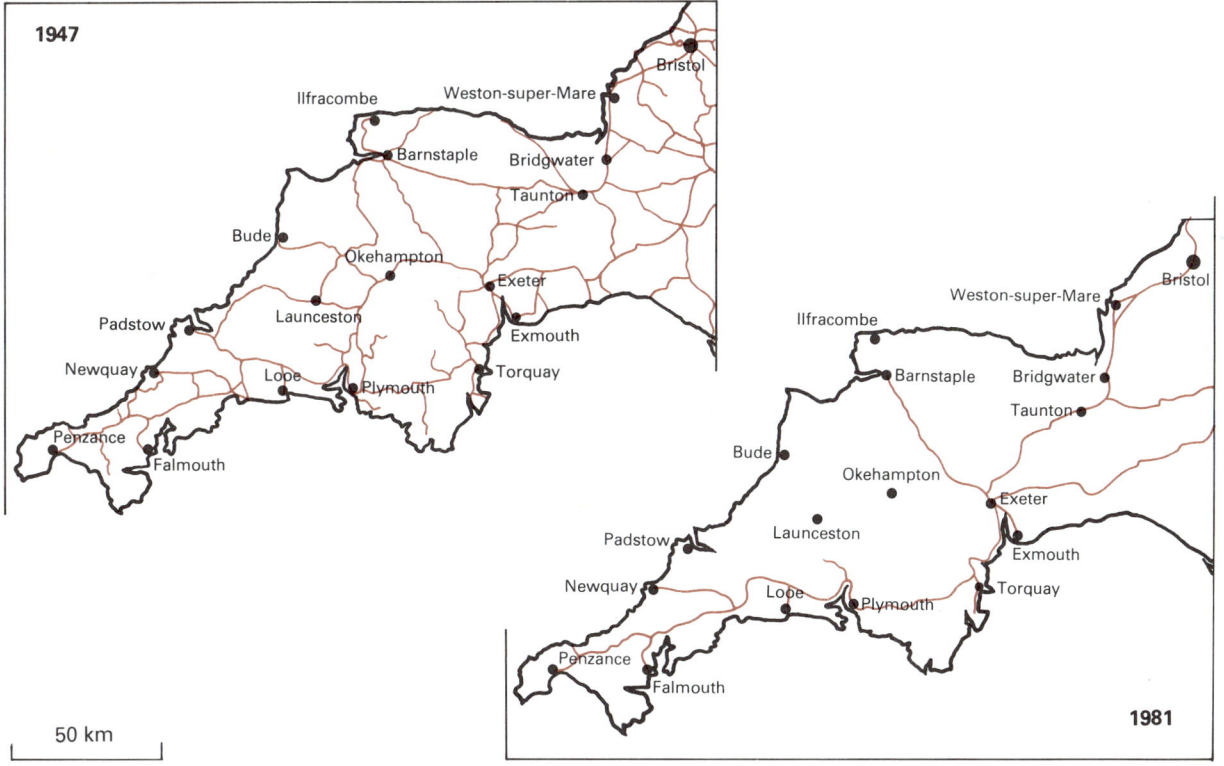

73

Fig. 91a A steam locomotive, now replaced by the familiar diesel shown in Fig. 91b

Fig. 91b A modern inter-city diesel engine

Fig. 92 A container is lifted from a train onto a waiting trailer at Felixstowe's Landguard Container Terminal. (See also Fig. 121.)

Fig. 93 British Rail's 200 km/h passenger train services began in 1976 when the first 'Inter-City 125' entered service between London and Bristol

congestion force many workers to go by train. Rush-hour services carry 2 million passengers per day. The fact that workers travel at limited times means that trains carry few passengers for most of the day. Despite efforts to encourage off-peak passengers by lower fares, and the fact that many people now work more flexible hours, the difference between the number of passengers travelling at peak times and the number travelling outside these hours still remains great.

Fig. 94 shows that, since 1954, there has been a steady decline in the percentage of passengers and

Fig. 94 Passenger and goods transport in Britain, 1954 to 1980

Passenger transport in Britain

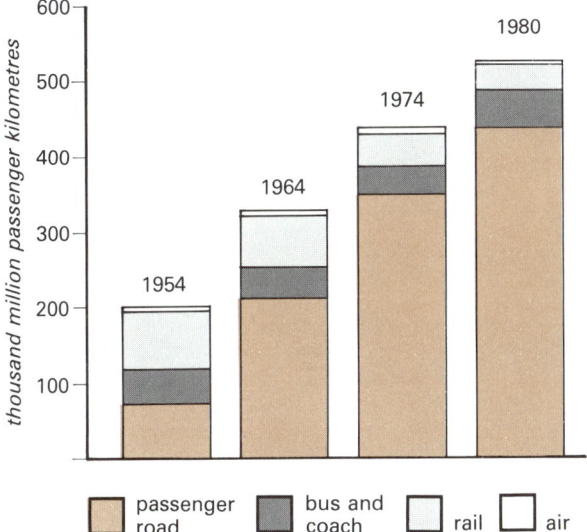

Goods transport in Britain

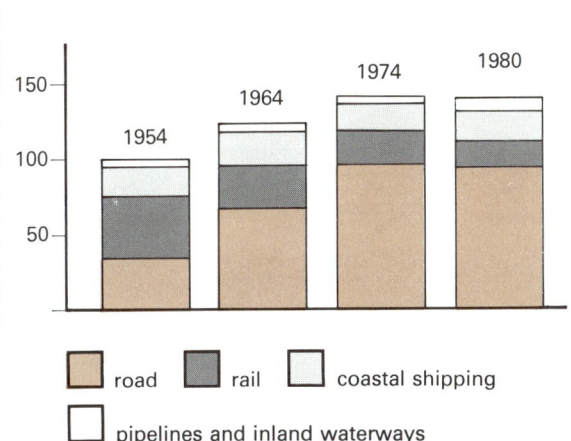

Exercise 34 Railways

1 Which one of the following describes 'a nationalised railway system'?
   (a) a nationwide railway network;
   (b) railways owned by a single company;
   (c) railways owned and controlled by the government.
2 Which one of the following describes an advantage that road transport has over railways?
   (a) faster over long distances;
   (b) able to carry heavy, bulky loads;
   (c) rapid speeds in urban areas;
   (d) door-to-door service.
3 What was the purpose of the Beeching Report?
4 Describe two effects of the Beeching Report.
5 To which one of the following are railways best suited:
   (a) short distance freight;
   (b) local rural services;
   (c) long distance inter-continental routes;
   (d) long distance inter-city services in Britain;
   (e) services linking major cities to holiday resorts?
6 Which of the following statements are true of electrified railways?
   (a) They use no fuel.
   (b) They have reduced the demand for oil.
   (c) They can use current generated from coal, oil, nuclear power or HEP.

Exercise 35 Railways today

Answer each of the following statements true or false.

1 The total length of railway track has increased since 1950.
2 Steam-engines have been replaced by diesel and electric locomotives.
3 Containers may be carried by rail, road and ship.
4 Many rural lines have closed.
5 Inter-city services are called urban railways.
6 Railway transport is safer than road transport.
7 Underground railways serve most cities in Britain.
8 Underground railways reduce traffic on roads.
9 Modern factories are built alongside main railway lines.
10 Railways offer the most convenient form of passenger transport.

goods carried by the railways. To compete successfully against other modes of transport, railways must attract the traffic for which they are best suited. Fast inter-city trains have advantages over air services in terms of price and time. Long-distance rail freight has performed less well against rival road haulage which offers door-to-door service with minimum loading delay. However, railways have an advantage over forms of transport which run on oil-based fuel. As oil becomes more expensive, railways will continue to run on electricity generated by coal, HEP or nuclear power.

Now try Exercises 34 and 35.

Road transport

Road transport is the most common method of carrying people and goods. Railways, ships and aircraft need special facilities for loading and delivery whereas cars give freedom of movement; few addresses in Britain lie beyond their reach. Specialised vehicles have been designed for particular needs. The bus, ambulance, ready-mix concrete lorry, containerised articulated lorry, mobile shop and motor-caravan all enjoy the freedom of the road.

Fig. 95 The increase in motor vehicles in Britain

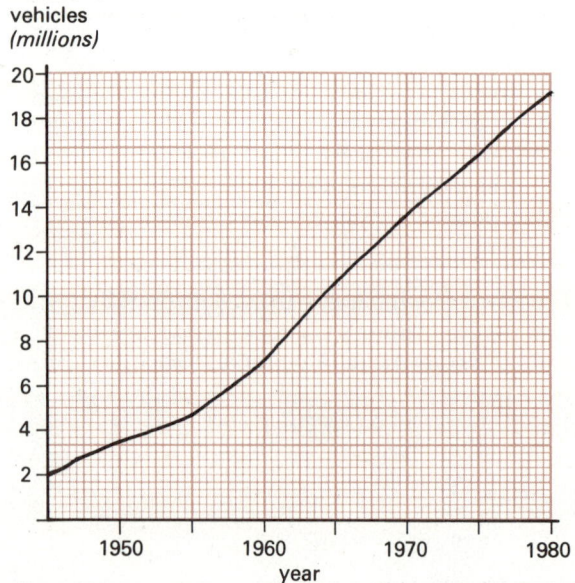

If the nineteenth century was the age of the train, then the twentieth century has been the age of the motor vehicle. Fig. 95 shows the remarkable increase in the use of motor vehicles since 1950.

Early roads in Britain

The Romans designed Britain's first road network nearly 2000 years ago (Fig. 96). They built straight roads which enabled soldiers to march quickly between forts and towns. Not until the motorway network of today was there another nationally-planned system of roads. Fig. 97 shows how the character of Britain's road network has changed since 1920.

Most of our roads follow ancient roundabout pathways between villages and towns. Until recently, there were few routes linking major towns. Since 1920, roads have been gradually widened, straightened and their surfaces improved, but road building has struggled to cope with traffic increases. Moreover, better roads may not solve the problem since improvements

Fig. 96 The road network in Roman Britain

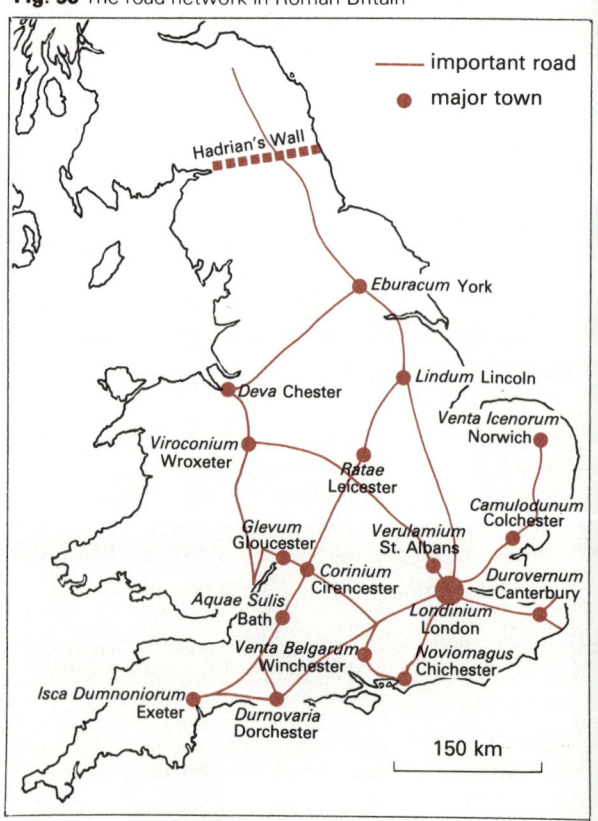

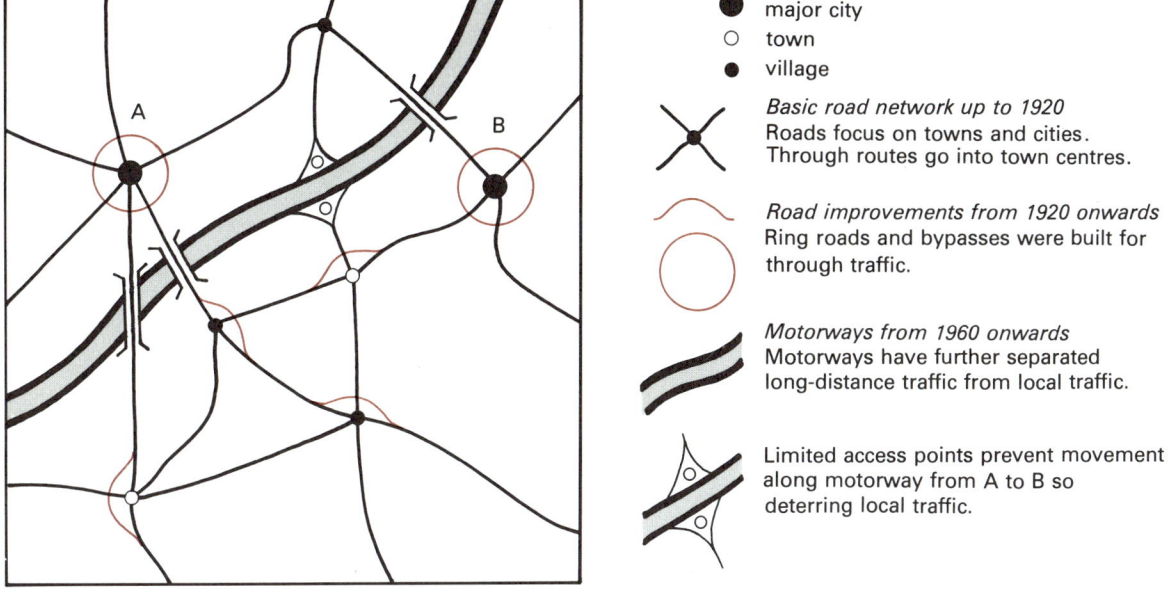

● major city
○ town
● village

Basic road network up to 1920
Roads focus on towns and cities.
Through routes go into town centres.

Road improvements from 1920 onwards
Ring roads and bypasses were built for
through traffic.

Motorways from 1960 onwards
Motorways have further separated
long-distance traffic from local traffic.

Limited access points prevent movement
along motorway from A to B so
deterring local traffic.

Fig. 97 A typical road network in Britain showing the successive stages of road building

encourage still more vehicles. If one stretch of road is widened, congestion increases at other bottle-necks. Rush-hour traffic has become a major problem for town planners.

Motorways

In 1953, a network of new motorways was planned following examples set by the Italian 'autostrada' and the German 'autobahn'. The first small section, the Preston bypass, was opened in 1958. Today, Britain's motorway network (Fig. 98) totals 2600 kilometres. Although this forms only a tiny part of the 360 000 kilometres of roadway in Britain, motorways carry most long-distance traffic including much of the heavy traffic between major industrial regions. Proximity to motorways is an important factor when planning new factories and housing.

Motorways have halved most journey times between cities. They are designed for fast long-distance traffic (Fig. 99) and offer several advantages over conventional roads. Pedestrians, horses and bicycles, for example, are banned. Access roads are limited in number and local roads pass beneath or above the motorway. Two or three lanes in each direction help traffic to move freely, while slip roads take vehicles to and from the motorway without interrupting the main traffic

flow. Gentle bends permit continuous high speeds, whilst cuttings and embankments reduce gradients to a minimum.

Safety measures help to reduce accidents. The edge of the motorway is usually kept free from houses or business premises, apart from service stations which offer fuel, food and other conveniences at suitable intervals. Steel barriers along the central reservation separate the traffic. An extra inside lane or 'hard shoulder' gives refuge for breakdowns and a passage for vehicles when other lanes are closed. Large signs give clear and early warnings to drivers, and telephones at regular intervals mean that rescue services can be called in an emergency.

In 1980, motorways cost between £4 million and £10 million per kilometre to build – depending upon building costs and the value of land needed. However, the cost of motorways cannot be measured just in terms of cash. A motorway needs a strip of land up to 80 metres wide, so many thousands of hectares of farm land are lost each time a motorway is built. Rural and urban communities living alongside motorways are often disrupted by the noise and air pollution caused by traffic. Although motorways have eased traffic flow between cities, delays inside the urban area continue to occur.

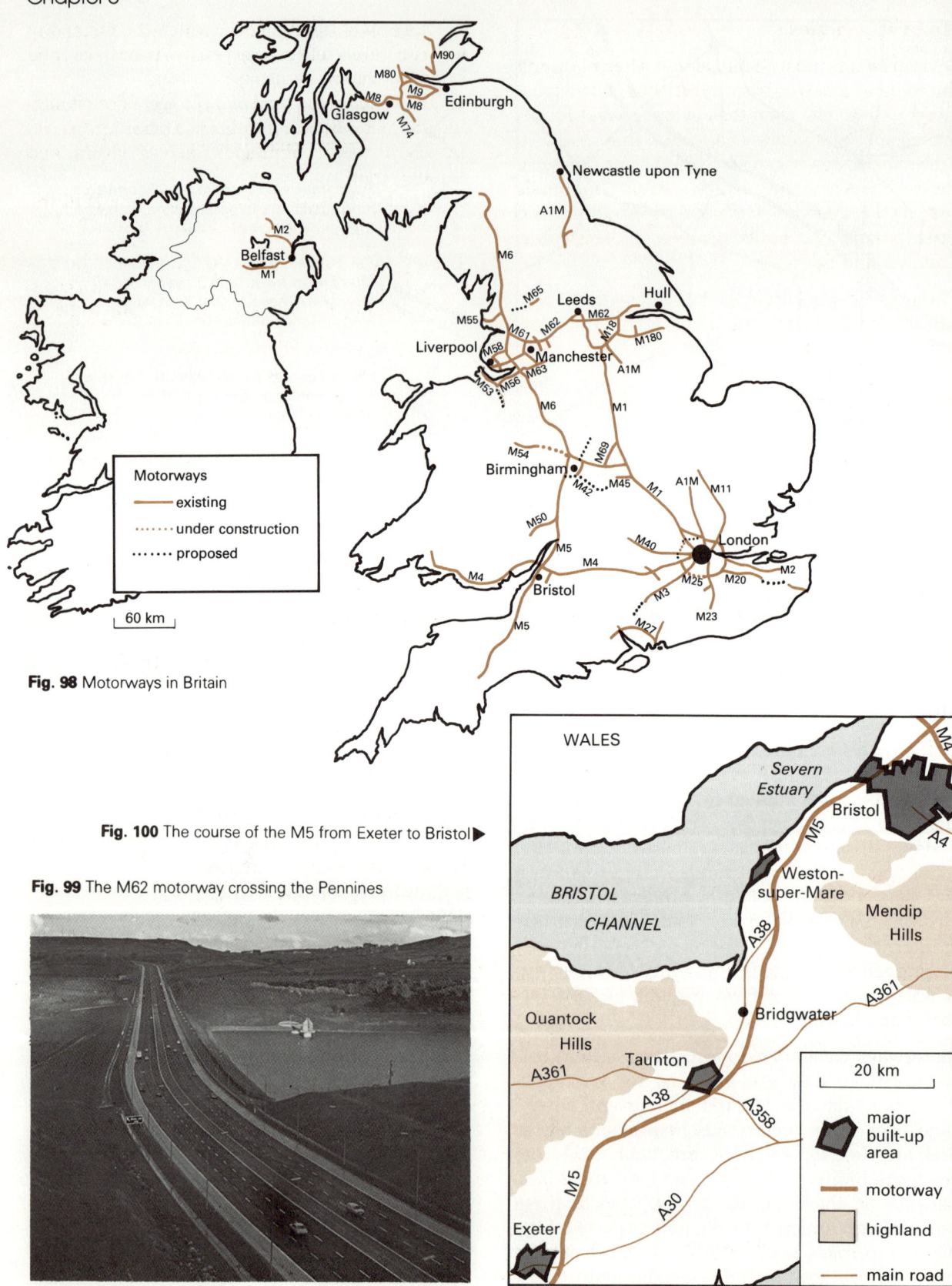

Fig. 98 Motorways in Britain

Fig. 100 The course of the M5 from Exeter to Bristol ▶

Fig. 99 The M62 motorway crossing the Pennines

Motorway routes

Several factors must be considered when planning the *route* of a new motorway. Although the most direct path might seem the obvious route to take, few motorways follow a straight line. Fig. 100 shows the course taken by the M5 from Exeter to Bristol. Note how the motorway avoids highlands such as the Quantock Hills and the Mendips, and skirts around the built-up areas of Bridgwater, Taunton and Exeter.

Planners have to consider the following factors when building a motorway:

1. The cost of the land over which the motorway will run. It is likely to be expensive near towns and in good farming areas.
2. The relief of the land and the cost of constructing the road – building costs are high where a road travels through highlands, across rivers and through built-up areas.
3. Possible damage to the environment. Will the motorway spoil an area of natural beauty?

The cheapest route in terms of the cost of land is often the most expensive to construct. Such a problem is explored in more detail in Exercise 38.

Now try Exercises 36, 37 and 38.

Exercise 36 Motorway service centres

1 Name four purposes served by motorway service centres.
2 Explain why it is necessary to have services on both sides of a motorway.
3 Name two services which might be provided on only one side.
4 Which of the following takes up most of the land area covered by the service centre:
(a) buildings; (b) grass verges; (c) access roads; (d) parking space?
5 From the following list name the features marked A, B, C, D and E on the photograph: bridge and restaurant; access road to motorway; slip road from motorway; central barrier; service road to centre.

6 Explain the difference between the use of area X from that of area Y.
7 Describe three problems faced by owners of the surrounding farm land as a result of the motorway.
8 A motorist in need of petrol fails to take the slip road at A. Which of these courses of action are safe and lawful:
(a) stop and reverse;
(b) cross to access road on other side;
(c) enter service station at B;
(d) continue journey to next service station;
(e) leave motorway at next exit;
(f) drive to hard shoulder and use emergency telephone?

Exercise 37 Which way?

A motorist wishes to travel from D to C. He can make the journey using 'A' class roads only, or by taking the motorway for part of the way. Take the average speed on 'A' class roads to be 50 km per hour and the average speed on motorways to be 100 km per hour.

The two journeys are compared in the table below.

1 Draw similar tables to calculate these journeys:
  B to A (i) direct by 'A' class road;
  (ii) using motorway from X to Y.

  D to B (i) travelling by 'A' class roads via A;
  (ii) using motorway from Y to X.

2 In which journey(s) does taking the motorway
  (a) shorten the distance travelled;
  (b) reduce the journey time?

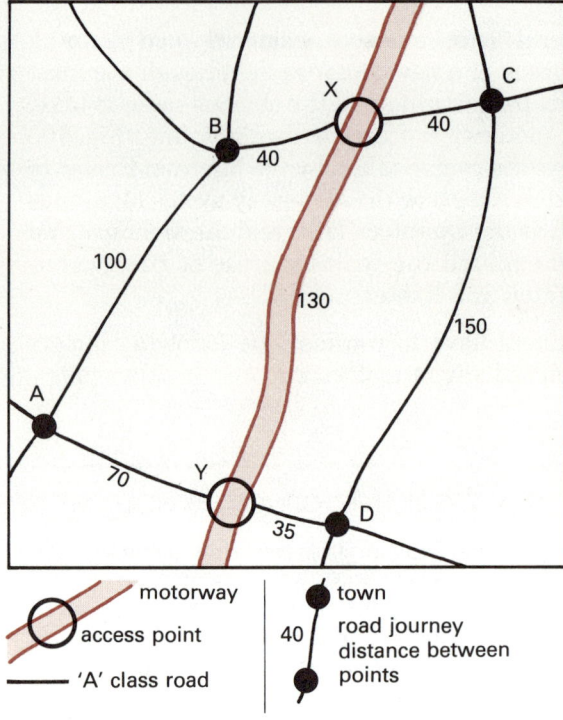

| | | motorway |
| access point |
| | 'A' class road | town |
| road journey distance between points |

Route	Distance (km)			Time		
D to C	'A' class roads	Motorway	Total	'A' class roads	Motorway	Total
'A' class road only	150	—	150	3h	—	3h
Via motorway and 'A' class road	75	130	205	1½h	1h 18min	2h 48min

Exercise 38 Building a motorway

The map shows an imaginary area and illustrates some of the problems faced by planners when deciding upon the course of a new motorway. Each hexagon contains a number which represents the cost of the land needed for the motorway. In addition, construction costs must be paid; they vary with the type of land and are given in the key. Note that some of the cheapest land, such as hills and marshland, is expensive to build on, as cuttings, embankments or drainage schemes are needed. The areas where building costs are lowest are often where land values are high, such as fertile farm land and urban zones.

Example

Two possible routes are proposed for a motorway to link Babham with Cabton. The land and building costs for each route have been calculated and are shown in the table. The shortest route is directly through the hills where total land costs are low (£12 million) but construction costs are high (£22 million), giving a total cost of £34 million. For route Y, building costs are much lower (£16 million), but land costs are higher because it is a longer route (£14 million). Although longer than X, route Y costs £4 million less in total than route X.

This example will help you to do the exercise on page 82.

Exercise 38 continued

Possible routes between
Babham and Cabton

Route	Cost (£m)		
	Land	Construction	TOTAL
X	12	22	34
Y	14	16	30

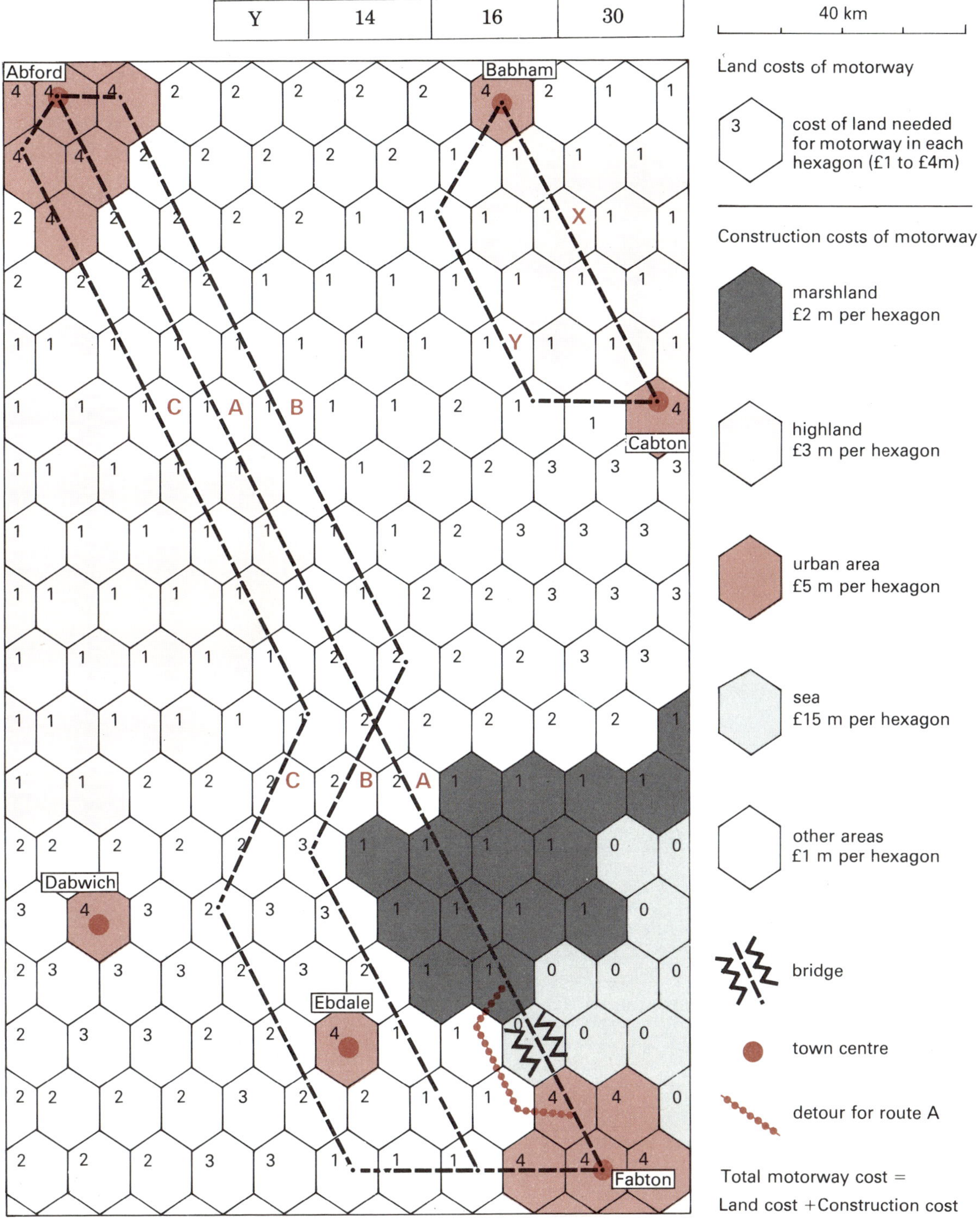

40 km

Land costs of motorway

3 — cost of land needed
for motorway in each
hexagon (£1 to £4m)

Construction costs of motorway

marshland
£2 m per hexagon

highland
£3 m per hexagon

urban area
£5 m per hexagon

sea
£15 m per hexagon

other areas
£1 m per hexagon

bridge

town centre

detour for route A

Total motorway cost =
Land cost + Construction cost

Exercise 38 continued

Study carefully the proposed routes A, B and C for a motorway to link Abford with Fabton. Remember that the total costs of completing the motorway include (a) the cost of the land, and (b) construction costs.

Route	Cost of land	Construction costs	Total costs
A			
B			
C			

1 Copy and complete this table.
2 Calculate the distance of each of the routes A, B and C.
3 The cost of the bridge on route A is now considered too high. Calculate the effect upon total costs of the detour (shown by the dotted line).
4 Calculate the total costs of building a motorway directly from: (a) Abford to Babham; (b) Babham to Dabwich; (c) Cabton to Ebdale.
5 Calculate the detour index for routes B, C, X and Y.
6 Give three reasons why, despite its higher cost, planners may prefer route X to route Y.

Road traffic in towns

Traffic congestion

Fig. 101 shows the typical pattern of roads found in many towns. Most major roads lead to the town centre. Such roads are known as *radial* routes. Fig. 102 illustrates the flow of vehicles along a radial road during a typical morning rush-hour.

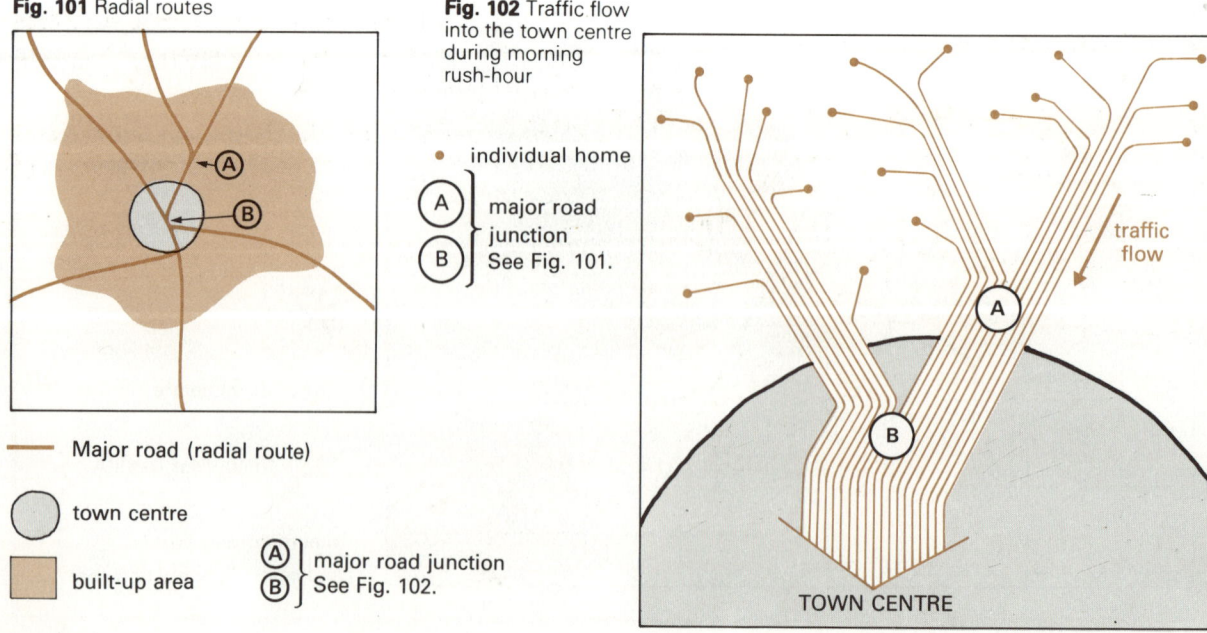

Fig. 101 Radial routes

— Major road (radial route)

◯ town centre

▨ built-up area

Ⓐ } major road junction
Ⓑ } See Fig. 102.

Fig. 102 Traffic flow into the town centre during morning rush-hour

• individual home

Ⓐ } major road
Ⓑ } junction See Fig. 101.

traffic flow

TOWN CENTRE

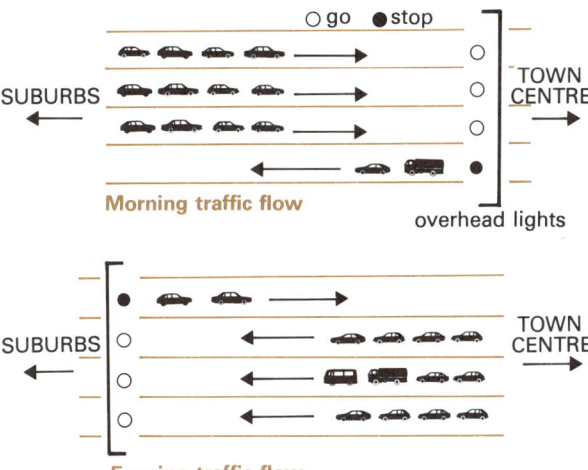

Fig. 103 Tidal traffic flow to relieve congestion during rush-hours

People leave their individual homes to catch buses or to drive themselves to work-places and shops in the town centre. As they move towards the centre they are joined by other travellers. At each road junction, the flow of traffic increases until people reach their destinations in the town centre. During the evening rush-hour, this pattern is reversed as workers make their way home.

As traffic converges upon the town centre, road space becomes increasingly crowded and so average vehicle speed is reduced. Traffic flow is also

Fig. 104 A typical bypass and ring road

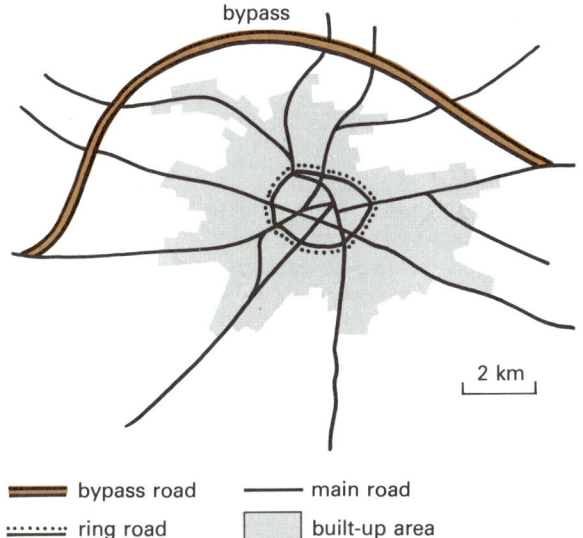

hindered by increasing numbers of road junctions, many of which are controlled by traffic lights. Traffic congestion in a London rush-hour can reduce vehicle speeds to less than 15 kilometres per hour in places. Eighty percent of commuters travelling into central London therefore use the Underground, surface railways or buses.

Improving traffic flow

To reduce traffic congestion, existing roads can be improved. Recent measures to increase traffic flow have included introducing the following:
1. one-way streets;
2. bus-only lanes;
3. no-parking clearways;
4. no right turns at major junctions (see Fig. 108);
5. tidal flow systems (see Fig. 103) which use overhead lights to reverse the direction of traffic in central lanes. For example, in a four-lane road, morning traffic uses three lanes into the city; evening traffic uses three lanes out.

Bypasses and ring roads

So great has been the increase in vehicle numbers since 1950 (Fig. 95, page 76), that new roads have become necessary. The *bypass* (Fig. 104) is designed for long-distance 'through traffic' which has no need to go to the town centre. Its route usually follows the edge of the built-up area where it connects with other long-distance main roads. In some cases the bypass may form a complete circuit around the town. The London *orbital* route (Fig. 105) is an example.

The inner *ring road* (Fig. 104) encircles the city centre and carries local traffic from one side of town to the other.

Now try Exercises 39 and 40.

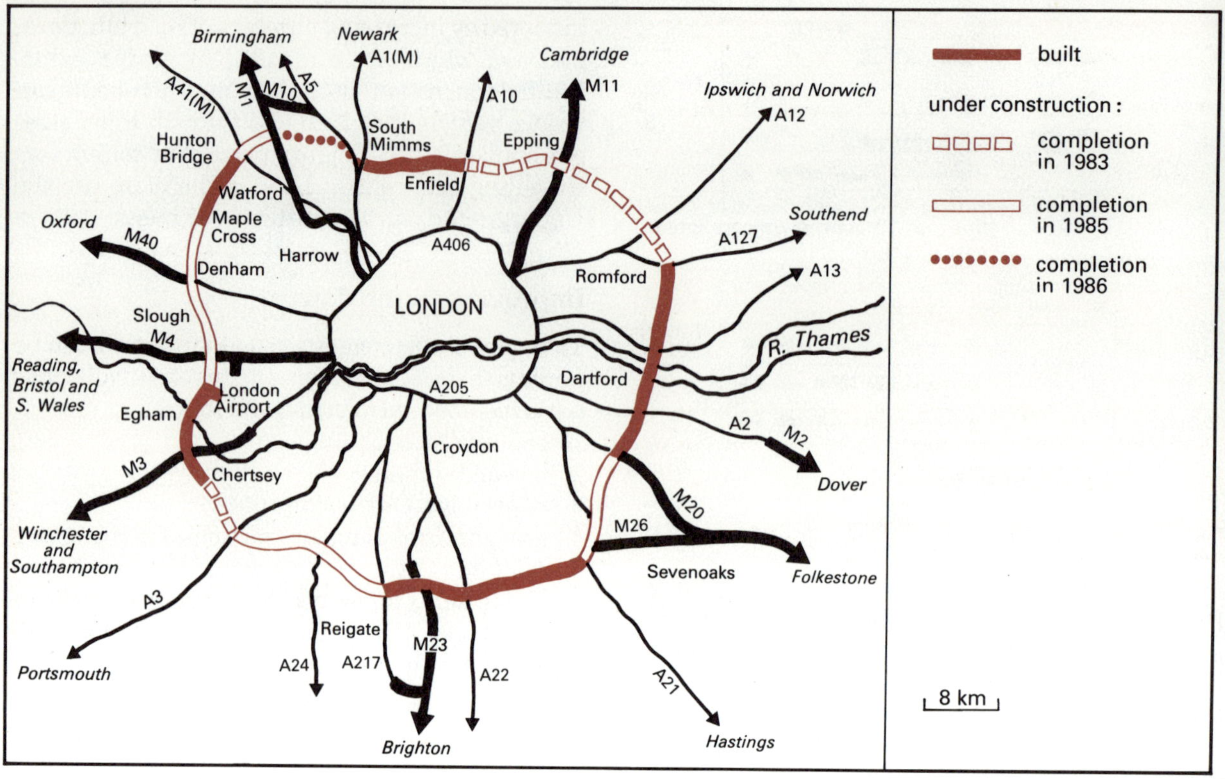

Fig. 105 London's orbital route

Exercise 39 Road junctions

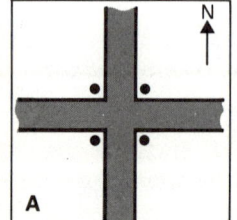
A

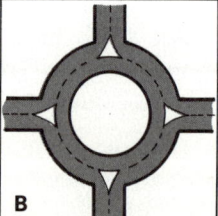
B

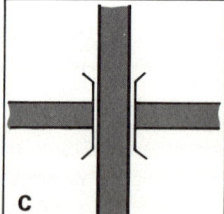
C

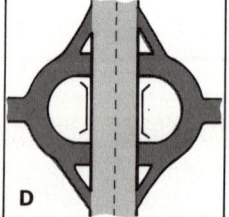
D

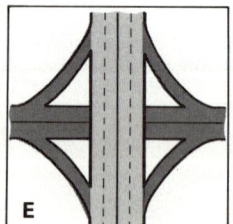
E

Road junctions frequently cause traffic delays. The diagram above shows five different road junctions:

1. Identify each type from the following descriptions:
   (a) road junction with traffic lights;
   (b) single-level roundabout with access to north, south, east and west;
   (c) motorway flyover with complete access to the motorway from the main road;
   (d) motorway flyover with restricted access to the motorway from the main road;
   (e) split-level underpass with no access.

2. At which junction(s) is it:
   (a) not possible for west-bound traffic to change course to eastwards;
   (b) not possible for west-bound traffic to change course to northwards;
   (c) not possible for south-bound traffic to change course to westwards;
   (d) not possible for south-bound traffic to change course to eastwards?
3. Which junction(s) need road bridges?
4. At which junction(s) do some vehicles have to stop?

Exercise 40 Road transport

1 Who built the first road network in Britain? What was its main purpose?
2 Describe one advantage and one disadvantage of a straight road.
3 Why were roads not improved to any great extent from 1850 to 1900?
4 Give one reason why road improvements do not always reduce traffic congestion.
5 Describe five ways in which motorways allow traffic to flow quickly.
6 What is an urban motorway?
7 Describe three disadvantages of urban motorways.
8 Give two reasons why there are no motorways in Cornwall.
9 Which of the following conditions causes most motorway accidents: ice; rain; wind; fog?

10 Match each of these road features with the correct definition.

(a) bypass	—a walking area for shoppers.
(b) pedestrian precinct	—a detour for through traffic.
(c) clearway	—a method of increasing speed of public transport.
(d) multi-storey car-park	—stops parking on main roads.
(e) lay-by	—reverses direction of vehicles in central road lanes.
(f) bus lane	—increases parking spaces in city centres.
(g) underpass	—causes circular flow of traffic.
(h) tidal traffic flow	—separates traffic into two levels.
(i) roundabout	—permits roadside parking without holding up traffic.
(j) no right turns	—improves traffic flow along central lanes at road junctions.

Measuring traffic flow

Before any road improvements are made, planners need an accurate picture of the volume of traffic. Information can be gathered by *traffic surveys*. Here are three simple kinds of survey.

1. A simple *traffic count* can be made of the number of vehicles passing a given point during a certain period of time. Fig. 106 shows the type of record form used in such a survey. Note how time is divided into convenient 20-minute intervals,

Fig. 106 Traffic count record form

TRAFFIC COUNT   Observer's Name.... W. Grant....
Date.... 4 April....
Location.... Outside Bellemoor Hotel, Hill Lane, Southam
Direction of traffic.... Southwards to City Centre....
Weather conditions.... Dull, raining....

Time from	Time to	Cars	Buses	Goods vehicles	Motor cycles	Total vehicles per 20 min
08·00	08·20	⩗⩗ ⩗⩗ ⩗⩗ ⩗⩗ 20	⩗⩗ ⩗⩗ ⩗⩗ ‖ 17	⩗⩗ ‖‖‖ 8	‖ 2	47
08·20	08·40	⩗⩗ ⩗⩗ ⩗⩗ ⩗⩗ ⩗⩗ ⩗⩗ ⩗⩗ ⩗⩗ ⩗⩗ ⩗⩗ ‖ 51	⩗⩗ ⩗⩗ ⩗⩗ 15	⩗⩗ ⩗⩗ ⩗⩗ 15	‖‖‖ 3	84
08·40	09·00	⩗⩗ ⩗⩗ ⩗⩗ ⩗⩗ ⩗⩗ ⩗⩗ ⩗⩗ ⩗⩗ ⩗⩗ ‖‖‖ 48	⩗⩗ ⩗⩗ ⩗⩗ ‖ 16	⩗⩗ ⩗⩗ ⩗⩗ ⩗⩗ ‖‖ 22	⩗⩗ ⩗⩗ ‖‖‖ 13	99
Column totals per vehicle type		119	48	45	18	Total vehicles per hour 230

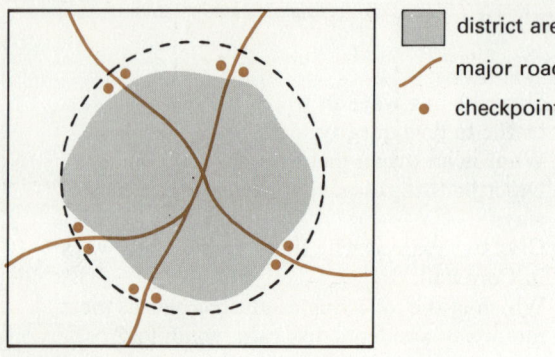

district area

major road

checkpoint

Fig. 107 Traffic checkpoint positions for a ring count

and vehicles classified into four broad groups. The intervals and vehicle classes can of course be modified to suit traffic circumstances and the purpose of the survey.

2. A *ring count* (Fig. 107) can be made of the number of vehicles entering and leaving a particular district. Checkpoints are placed at suitable points on each side of the radial route to be surveyed. The total number of vehicles leaving and entering the ringed zone can therefore be observed and recorded. This gives a useful idea of the traffic

Fig. 108 Traffic flow at a typical road junction

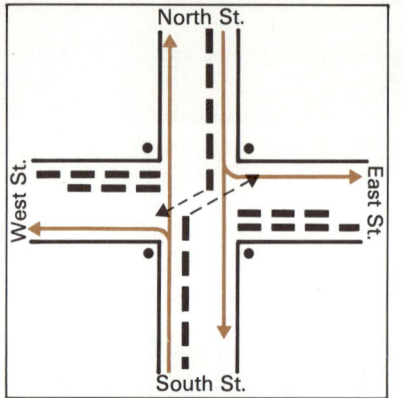

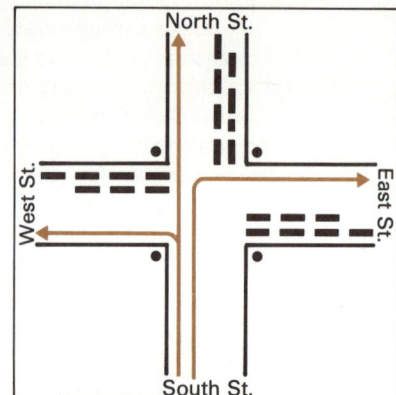

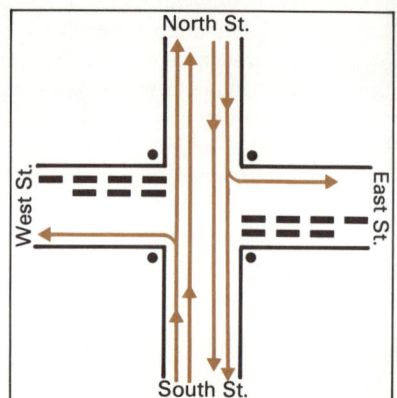

a Lights showing 'go' signal to traffic from North St. and South St. *All* right turns allowed.

b Lights showing 'go' signal to traffic from South St. only. Right turn allowed.

c Lights showing 'go' signal to traffic from North St. and South St. *No* right turns allowed.

▬ ▬ ▬ stationary vehicles   ➝ moving traffic   ● traffic light   ‑ ‑ ‑ ➤ desired path blocked

Fig. 109 Junction count record form

JUNCTION COUNT	Observer's name W. Grant

Date 5 May

Location JUNCTION OF EAST ST, WEST ST, NORTH ST, SOUTH ST.

Weather Conditions Dry and Sunny

Traffic from WEST STREET.

Time		INTO			
from	to	NORTH STREET Number of vehicles	EAST STREET Number of vehicles	SOUTH STREET Number of vehicles	Total vehicles
08·00	08·05	ᚷ 5	ᚷ 5	‖ 2	12
08·05	08·10	ᚷ ᚷ 10	ᚷ ᚷ 10	‖‖ 3	23
08·10	08·15	ᚷ ᚷ ᚷ ᚷ ᚷ 25	ᚷ ᚷ ᚷ 15	ᚷ 5	45
Total vehicles		40	30	10	80

volume in the district. Such a survey is often used when a bypass for a town or village is being considered.

3. A *junction count* can be done at busy road junctions. Traffic lights and vehicles waiting to turn right can cause serious delays (Fig. 108). A junction count may provide information to justify, for example, retiming the traffic lights or introducing a one-way system. Fig. 109 shows the type of form used for recording a junction count. Four such counts are needed to record traffic movements from each of the four roads at the junction. Fig. 110 illustrates how the results of such a survey can be shown.

Now try Exercises 41, 42 and 43.

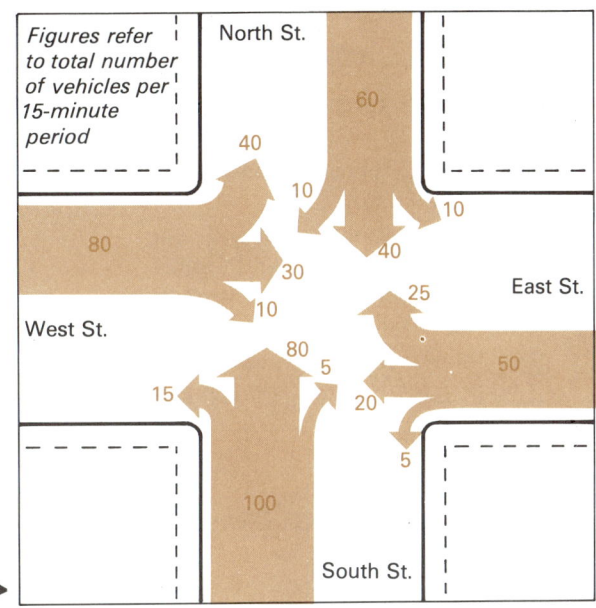

Fig. 110 Recording the results of a junction count ▶

Exercise 41 London traffic

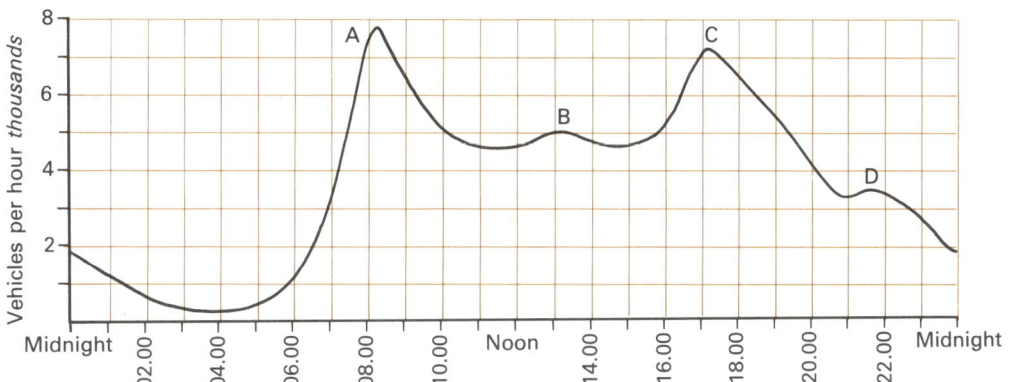

The graph plots the results of a traffic count taken as vehicles travel in and out of the central London area.

1 In which direction will most traffic be moving at A—into or out of London?
2 Choose the correct description for each of points A, B, C and D from the following:
   morning rush-hour; workers returning home; lunch-hour; visitors to clubs and cinemas.
3 At what time is traffic flow (a) heaviest; (b) lightest?
4 Which of the following reasons explains why peak C is lower than peak A?

(a) Some workers stay in central London for evening entertainment.
(b) Some shoppers return home before midday.
(c) Many people travel home at 17.00.
(d) People start work between 08.00 and 09.00 but leave work at times between 16.00 and 19.00.
5 Which of these changes would reduce traffic flow in central London:
   (a) cheaper rail fares;
   (b) increased Underground fares;
   (c) increased parking-meter charges;
   (d) improved bus services;
   (e) cheaper petrol?

Exercise 42 Junction count

Study Fig. 110 carefully and answer the following questions.

During the 15-minute period of the survey . . .

1 How many vehicles:
   (a) turned north from West Street;
   (b) turned west from North Street;
   (c) travelled south along North Street;
   (d) travelled north along South Street?

2 How many vehicles travelled towards the junction in total? How many travelled away?

3 Look at the total two-way traffic for each road.
   (a) Which road is busiest? Give its total two-way traffic volume.
   (b) Which road is least busy? Give its total two-way traffic volume.

4 If one 'no right turn' sign could be installed, which position would: (a) cause least interference with traffic; (b) cause most interference? Select from: (i) facing South Street; (ii) facing North Street; (iii) facing East Street; (iv) facing West Street.

5 If the road junction in Fig. 110 is 2 kilometres from the town centre, in which direction is the town centre likely to be?

Exercise 43 Traffic flow

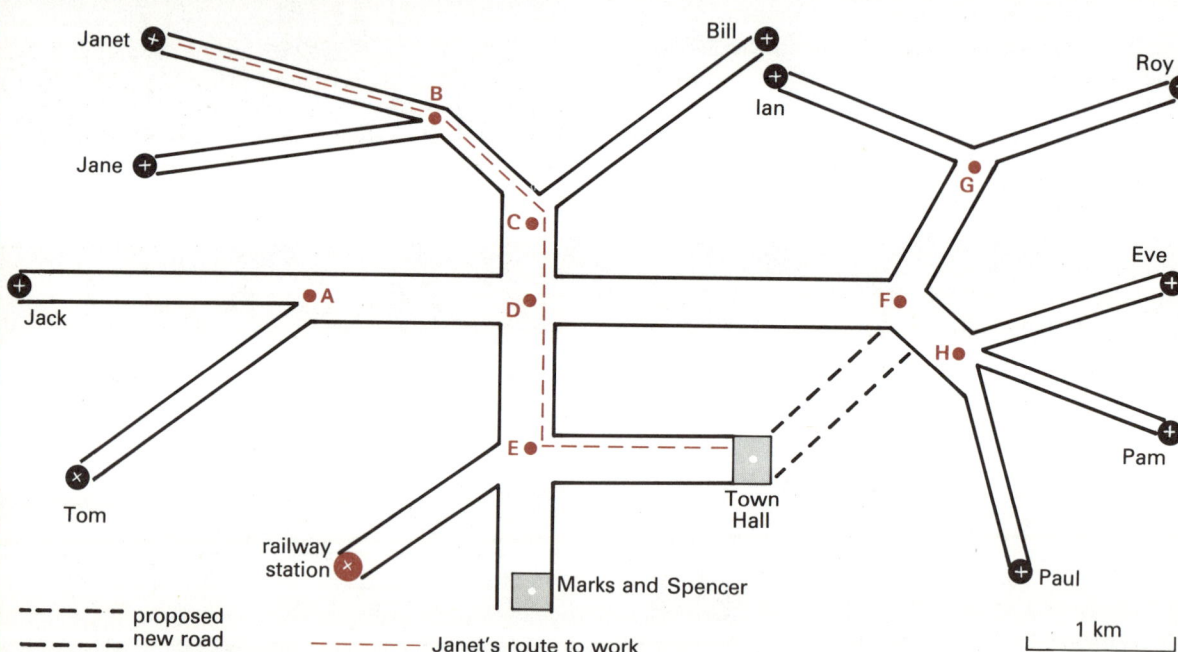

The people whose homes are located on the map make daily journeys to work. Tom, Jane, Janet and Paul work at the Town Hall; Jack, Ian and Eve work at the Railway Station; and Pam, Roy and Bill work at Marks and Spencer's. Janet's journey to work has been drawn.

1 Trace the map and complete a *traffic flow line map* by marking each journey to work.

2 Assuming each person travels alone by car, how many vehicles pass junction A; B; D; E?

3 Arrange the following road sectors in order of traffic volume. Put the busiest stretch of road first: F-D; H-F; G-F; D-E.

4 How many vehicles (a) turn right at junction F; (b) turn left at junction E; (c) turn right at junction D?

5 Who makes the longest journey? How long is it?

6 Who makes the shortest journey? How far is it?

7 Calculate the journey detour index for (a) Bill; (b) Paul.

8 A new road is planned to link the Town Hall with junction F. If the road is built . . .
   (a) who will use the road?
   (b) how many vehicles will then use road D-F?
   (c) what effect would such a road have on
      (i) the total length of Paul's journey;
      (ii) the detour index of Paul's journey?

Water transport

Water provides one of the cheapest means of transport. Water gives boats buoyancy and low friction so they can move easily. In early times, rivers carried people and goods in vessels such as the dug-out canoe. Present-day vessels, such as the giant barge-train seen in Fig. 111, are capable of carrying up to 100 000-tonne cargoes in twenty 5000-tonne barges.

Rivers

In Europe and North America, rivers have been developed as important routeways for the densely populated areas. The river Rhine, for example, mapped in Fig. 112, is the most important inland waterway in Europe. Rising in the Alpine mountains of Switzerland, it turns northwards from Basel to flow 800 kilometres to enter the North Sea at Rotterdam, the world's busiest port. The Rhine gives an uninterrupted, ice-free route upstream to Basel for barges up to 1000 tonnes, whilst vessels of up to 20 000 tonnes can reach Duisburg in the Ruhr. This cheap bulk transport has encouraged the growth of manufacturing along the banks of the river. The coal, steel, engineering and chemical works of the Ruhr north of Cologne help to make the area one of the world's greatest industrial regions.

Many of the world's longest rivers, however, remain undeveloped. The direction of flow of some rivers means that they are of no use commercially. The Russian rivers Ob, Lena and Yenisey, and the huge McKenzie river in Canada, flow northwards to the wastelands of the Arctic. Their value is further reduced by winter freezing and spring floods. Other rivers such as the Amazon and the Congo flow through undeveloped regions which generate little trade. Rivers flowing through areas of seasonal rainfall, such as the river Nile, may have too little water in dry seasons to be navigable. The severely winding course of some rivers, for example the Mississippi, greatly lengthens journeys and increases time and cost. Moreover, few rivers remain naturally navigable and must be dredged to remove clogging sand and mud.

Fig. 111 A giant barge train on the Ohio River, USA

Fig. 112 The Rhine waterway

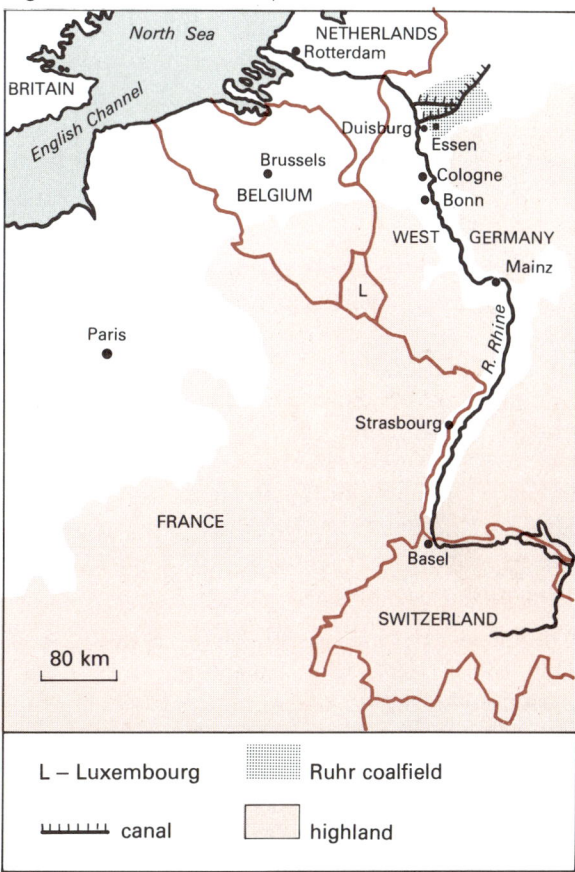

L – Luxembourg   Ruhr coalfield

canal   highland

Chapter 3

Canals

Many problems of navigation presented by rivers are overcome by canals. These vary in size from narrow barge canals such as those built in Britain between 1760 and 1830 to the Suez and Panama ship canals.

In Britain, during the early years of the Industrial Revolution, heavy goods such as coal and iron were carried by canal. This was an improvement upon cumbersome horse-drawn carts which had previously carried heavy goods. By 1830, more than 6000 kilometres of canals had been constructed. By 1880, however, the canals had lost most of their business to the much swifter steam railway. Canals are slow; to keep a level course they were often built to follow a winding route along a contour line; hills are crossed by awkward locks; delays often occurred when horse-drawn barges met. Today, most canals lie derelict, although some provide useful recreation facilities for pleasure boats and fishermen.

Fig. 113 shows the important inland waterways of Britain. Canals still carry heavy, bulky goods such as coal, steel and timber between Humberside and the towns of Leeds, Doncaster and Rotherham. Fig. 114 shows a barge train taking coal to a power-station along the Sheffield and South Yorkshire Navigation. Note the large width of the waterway compared with most barge canals in Britain.

In 1973, Britain joined the European Economic Community (the EEC or 'Common Market'). This boosted trade for North Sea ports and prompted the introduction of the Barges Aboard Catamaran system (BACAT). Specially-equipped twin-hulled vessels called catamarans (Fig. 115) sail across the North Sea with several barges locked in place between the two hulls. Regular services link Humberside with the Rhine so that barges can proceed directly between Leeds or Doncaster and Duisburg or Cologne.

The largest British canal is pictured in Fig. 116. Opened in 1894, the Manchester Ship Canal allows vessels up to 12 500 tonnes to reach Manchester, which is 50 kilometres from the Mersey estuary. Its course, mapped in Fig. 117, is lined with chemical works, food processing factories and power-stations which have been attracted by the cheap transport offered by the waterway. How-ever, because of the increase in the average size of merchant vessels, and improved motorway connections in the region, the Manchester Ship Canal faces an uncertain future.

Now try Exercise 44.

Fig. 113 Britain's major inland waterways

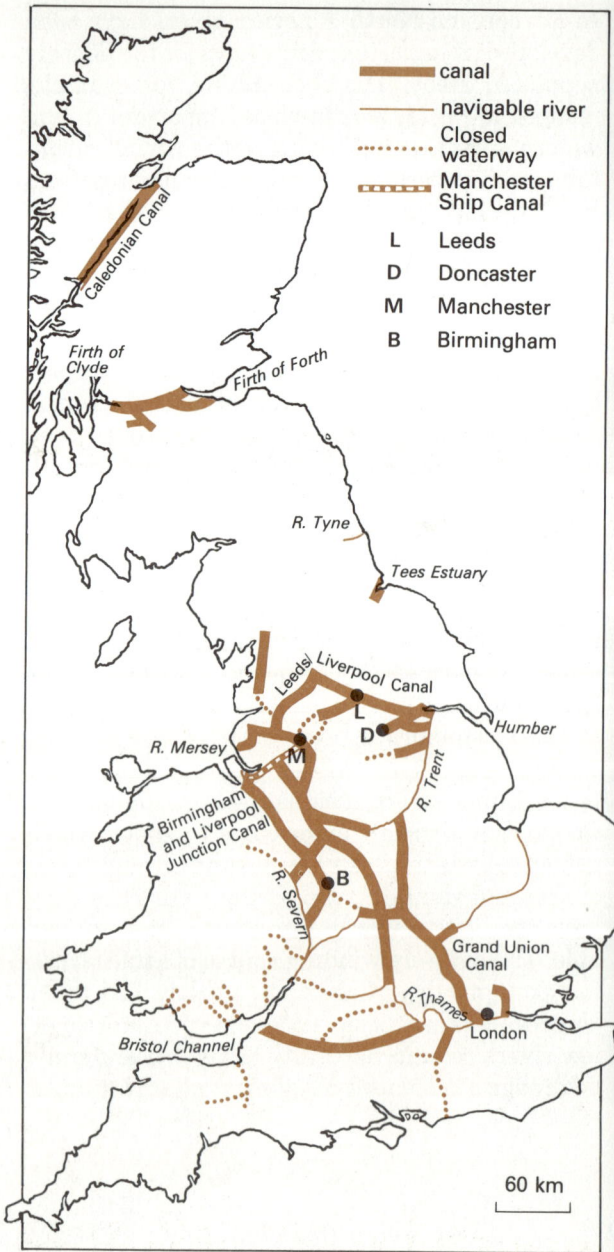

Fig. 114 Coal being taken to a power-station on the Sheffield and South Yorkshire Navigation

Fig. 115 The BACAT vessel being loaded carries ten 140-tonne barges on its main deck and three 70-tonne lighters between its twin hulls

Fig. 116 Part of the Manchester Ship Canal which changed the city from a river port to a sea port

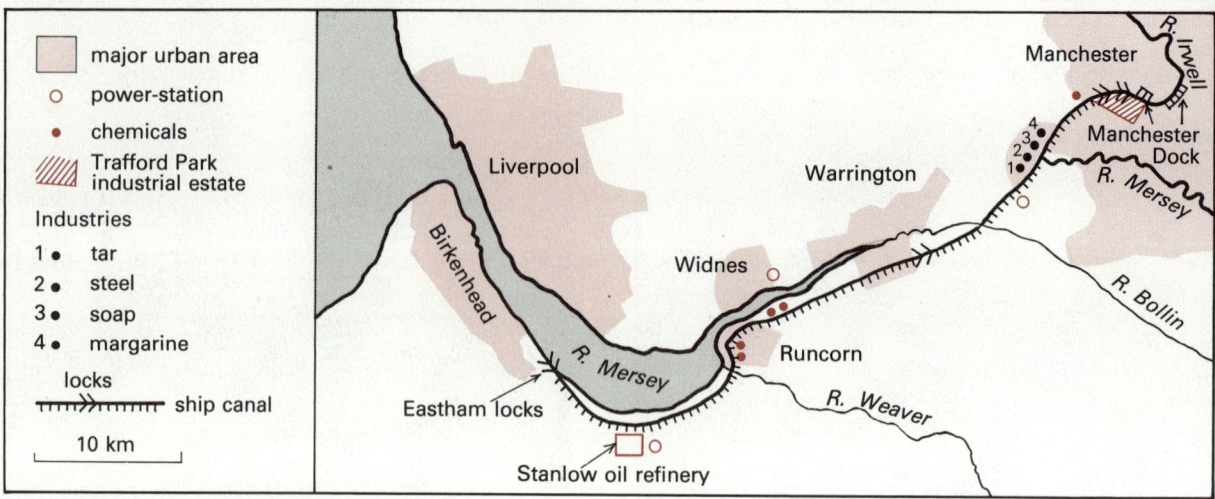

Fig. 117 The course of the Manchester Ship Canal

Exercise 44 Water transport

1 Place the following in historical order: nuclear submarine, dugout canoe, sailing-ship, horse-drawn barge, paddle steamer, turbine-driven liner, hovercraft.
2 Describe two advantages of water transport.
3 Describe two disadvantages of water transport.
4 Name the most important inland waterway in Europe.
5 Between which towns is the River Rhine navigable for 1000-tonne barges?
6 What is the Ruhr?
7 Describe three reasons why navigation is difficult on the River McKenzie in Canada.
8 Name one obstacle to navigation on (a) the River Nile; (b) the Mississippi; (c) the Ob.
9 Name three examples of ship canals.
10 Describe three disadvantages of barge canals.
11 Why are so many barge canals in England unused?
12 For which of the following cargoes would barge-canal transport be suitable: mail; newspapers; clay; coal; milk; sand?
13 What are the advantages of the BACAT system?
14 Name the largest British ship canal.
15 Describe four different kinds of industry located alongside the Manchester Ship Canal.

Ocean shipping

Ocean vessels offer the cheapest form of bulk transport. In recent years the average size of vessels has increased rapidly. In 1980, the average deadweight size of British vessels over 500 tonnes was 35 000 tonnes compared with 18 000 tonnes in 1970. In 1980, the British merchant fleet had 122 vessels with deadweight tonnage of more than 100 000 tonnes; twice the number in 1970. The larger the ship, the cheaper the cost of transport per tonne. Great savings are gained in terms of fuel, maintenance and crew. The cost of carrying each tonne of oil in a 250 000-tonne tanker is less than one-third of that charged by a 20 000-tonne vessel. Few ports, however, can take such huge ships and docks and harbours are costly to construct and maintain. Supertankers over 300 000 tonnes are unable to dock in any British port and must unload their cargoes in Bantry Bay in south-west Eire, from where smaller tankers complete delivery to Britain (Fig. 118).

Liners, tramps and bulk-carriers

Ocean transport is vital for industrial islands such as Britain and Japan. Over 99% by weight (about 75% by value) of Britain's overseas trade goes by sea. Ocean vessels may be grouped into three types.

1. *Liners* work regular timetables and carry passengers and goods. Severe competition from aircraft, however, has greatly reduced passenger shipping, especially over long-distance routes

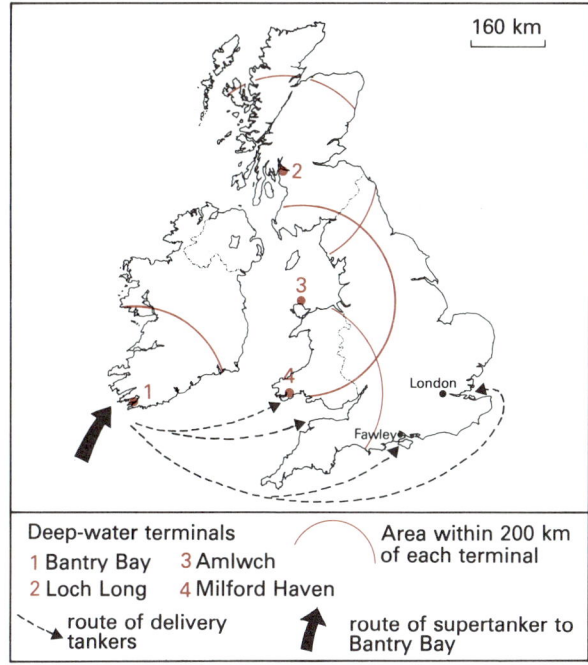

Fig. 118 The deep-water terminals supplying the British Isles

Fig. 119 An example of a tramp cargo route

such as the North Atlantic. The proportion of passengers travelling to or from Britain by sea is now about one-third compared with about one-half in the 1960s. Short-distance ferries, for example, the cross-Channel services, continue to survive. Freight-liners serve traders with small but regular consignments. Consequently, their total cargo includes a mixture of raw materials and finished products.

2. *Tramp* cargo ships do not follow regular routes or timetables. They sail wherever business is available. Plans for their journeys are arranged by their owners at a special market in London called the Baltic Exchange. The following voyage is typical: agricultural machines are carried from Liverpool to Lisbon, Portugal; wine and fruit from Lisbon to Stockholm, Sweden; timber and paper from Oslo, Norway, to Hull. The route is mapped in Fig. 119.

3. Specialist ships designed for one type of cargo now transport goods such as oil, iron-ore, coal and grain. Ships of this type are called *bulk-carriers*. An oil tanker, for example, has strict safety precautions to avoid fire or spillage. Bulk-carriers are loaded and emptied by mechanical means to reduce costly delays in port. Fig. 120 shows a bulk-carrier discharging iron-ore at the deep-water berth at Port Talbot, South Wales. The steelworks can be seen in the background.

Fig. 120 A 100 000-tonne bulk-carrier discharging iron-ore at Port Talbot

Containers and roll-on/roll-off

Until the 1940s, cargoes were loaded by cranes through hatches in the ship's deck. This slow and costly method needed many workers, and cargoes had to be split up into manageable packages. Today, the handling of cargo has been transformed by the *container*. This is a large metal box with an average volume of about 35 cubic metres. Goods weighing up to 25 tonnes can be packed inside strapped to wooden platforms called *pallets* which are easily stacked by fork-lift trucks. Containers are filled and sealed at the factory and, because of their regular size, they are quickly transferred between lorry, train and ship (Fig. 121). The most modern container ships are able to carry 3000 containers. All major ports are now equipped to handle containers, so costly delays are cut to a minimum.

Cargo handling has been streamlined on short sea journeys by the roll-on/roll-off (RO/RO) service. Some ships are built so that lorries can drive directly on board and off again at their destination. In other cases, articulated vehicles leave only the trailer on board. This is collected by another driving cab at its destination. Services have expanded where crossings are short, such as the North Sea routes to Europe from Harwich, Felixstowe and Hull; and the cross-channel services to France.

Now try Exercises 45 and 46.

Fig. 121
a A modern container ship being loaded at Felixstowe

b One of Felixstowe's 35-tonne capacity gantry cranes loading a lorry. (See also Fig. 92.)

Exercise 45 Ocean shipping

Copy and complete the passage using the ten correct words chosen from the list.

unload, petrol, liners, collect, Bantry Bay, cheapest, oil, costs, pounds, tonnes, dock, Fawley, terminals, supertankers.

Ocean tankers provide the ———— means of transporting ————. Vessels have been increased in size to reduce transport ———— per tonne. Special deep-water ———— are needed to accommodate ———— of over 300 000 ————. Because such huge vessels cannot ———— at ———— or even Milford Haven, they must ———— their cargoes at ————, from where smaller tankers can deliver oil to British ports.

Exercise 46 Ocean cargoes

Match the correct description to each of the following methods of sea transport:

(a) liners — an ocean-going vessel which carries canal barges.
(b) cross-channel ferries — can be carried by road, rail and sea transport.
(c) tramp cargo ships — carry lorries loaded with goods across the Channel.
(d) bulk-carriers — passenger ships severely hit by air transport since 1950.
(e) freightliners — large vessels built for carrying a specific cargo such as wheat or oil.
(f) containers — their route, timetable and cargo varies according to the business available.
(g) roll-on/roll-off ships — offer a shuttle service to the continent.
(h) BACAT barge — ships carrying assorted cargo over regular long distance routes.

Air transport

Aircraft provide the quickest means of transport over long distances. Over short routes, however, total journey times by air can be greater than by road or rail.

Since the first flight of 1903, aircraft development has been remarkable. Planes such as Concorde and the jumbo jet demonstrate the enormous progress made by twentieth century science and engineering. Supersonic Concorde can fly 100 passengers from London to New York in less than three hours whilst the Boeing Jumbo 747, with a payload of 500 economy passengers, has made international air travel available to millions of holiday-makers and business people.

Although air freight carries less than 1% of world trade by weight, it is of increasing importance. London's Heathrow airport, for example, handles 15% of the nation's total foreign trade by value. Goods such as precious stones, films, mail and newspapers, and perishable items such as flowers, obviously benefit from air freight services, but the range of items now includes medicines, scientific equipment and urgently-needed spare parts for cars and computers.

For the passenger, speed is the great advantage of air travel. Great savings in time can be achieved over long distances, an important consideration for people such as tourists and business people. Once outside the strict controls of air traffic regulations, an aircraft is free to take the most direct route. Services between London and Tokyo fly over the North Pole to save time and fuel. London to Tokyo by plane takes less than 24 hours, whereas a similar journey by sea takes four weeks. Tourists taking a two-week holiday in southern Spain spend a total of only six hours in the air and so get more time in the sun – going by road or rail would cut up to five days from the time they could spend at their resort.

In many regions, aircraft are the only suitable means of travel. Remote cattle-stations in Australia depend on flying doctors; isolated mining camps, trading-posts, weather-stations and defence bases owe their existence to air communications. Closer to home, air services from the Scottish mainland to offshore islands and oil rigs have done much to reduce their isolation (Fig. 122).

Fig. 122 A helicopter landing on a production platform in the Forties oilfield

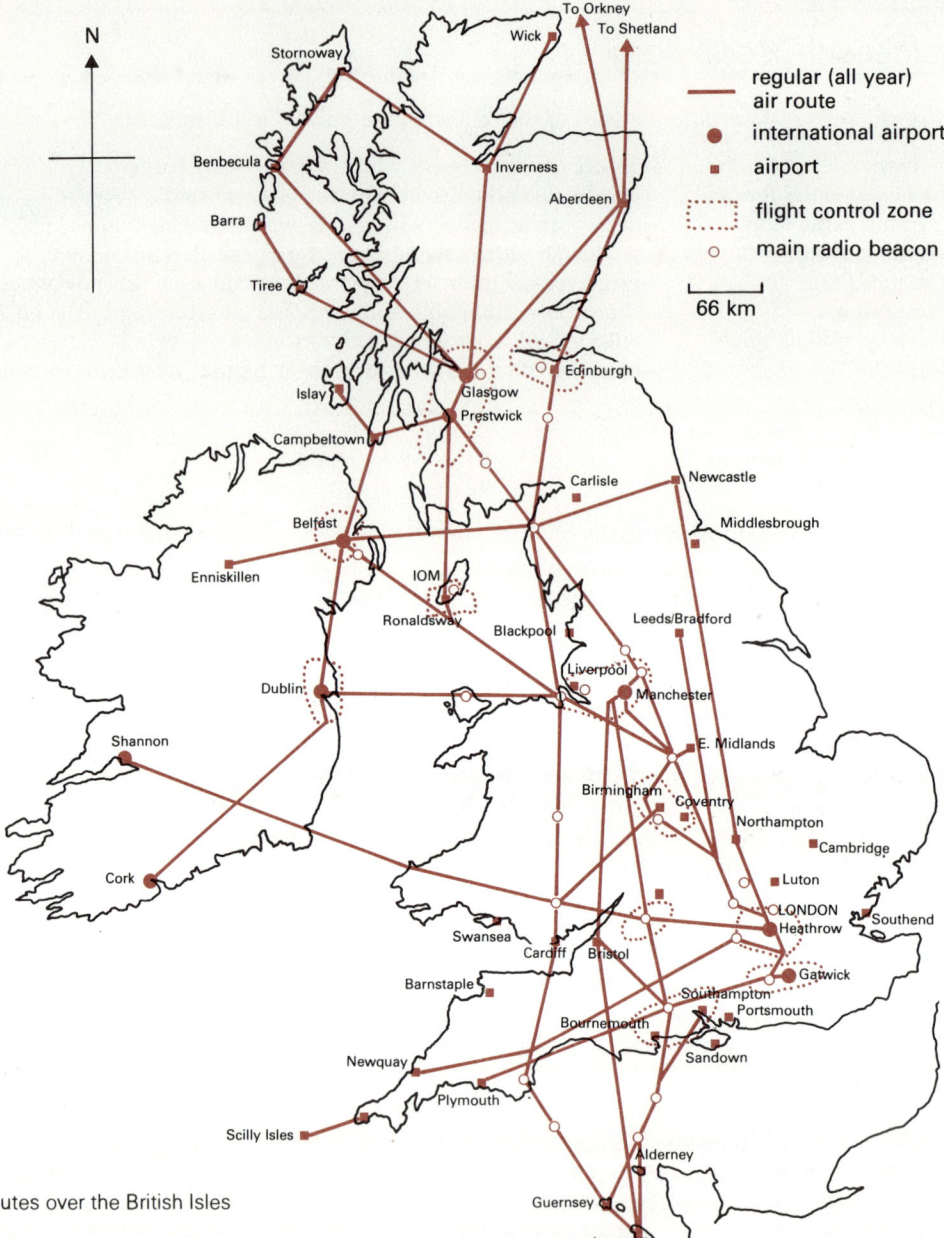

Fig. 123 Regular air routes over the British Isles

One major drawback of air transport is the higher cost. Passenger fares average four times those of rail, while charges for cargo may be up to a hundred times those of rail or sea. Aircraft are complex and valuable machines able to carry only small payloads. Design improves so fast that planes soon become out of date and must be replaced before their planned working life is through. Fuel bills, repairs and maintenance charges, together with highly-paid staff, bring high operating costs on top of the large investment needed to buy aircraft.

Another disadvantage is that air services may be too far away for many passengers to make use of them. As a result, road and rail services generally offer greater convenience than internal flights, especially as relatively short distances are involved. Once off the ground, air travel is speedy, but ground delays, getting to and from the airport, and waiting for good weather make many journeys much longer than expected. Internal air transport therefore remains little used in Britain. By contrast, the large distances that have to be travelled between cities in the USA make internal

Fig. 124 London's Heathrow Airport

air services much more viable and frequent scheduled flights make flying a popular means of travel. Fig. 123 shows the air services which operate in Britain.

Airports

International airports need plenty of space. Modern jets need runways over 3·5 kilometres in length. Heathrow's passenger and freight terminals, aircraft service areas, runways and carparks cover more than 10 square kilometres of what was once good farm land (Fig. 124). To be able to accommodate such large complexes, many airports are situated far from city centres. Passengers therefore waste a great deal of time reaching them. Heathrow, for example, is 20 kilometres west of central London.

London Heathrow is one of the world's busiest airports, employing more than 50 000 workers and handling over 25 million passengers and 500 000 tonnes of freight each year. The steady increase in traffic brings congestion in the air and on the ground. Aircraft queue in a spiral-shaped pattern above south-east England awaiting permission from air traffic control to land. Safety limits are stretched even with the use of computers and electronic guidance. To relieve congestion, a fourth terminal is planned at Heathrow and second terminals at Gatwick and Birmingham are possibilities. Airports in north-east Britain are being expanded due to the development of North Sea oil. Aberdeen now has one of the world's largest heliports which was opened in 1977 to service the North Sea rigs.

Now try Exercise 47.

Exercise 47 Air transport

A Are the following statements true or false?
1 Air transport is the fastest means of travel over long distances.
2 Concorde can fly faster than sound.
3 Air transport is the most common form of passenger travel in Britain.
4 Low-weight, high-value freight is not suited to air transport.
5 The introduction of inter-city rail services has boosted air passenger traffic from Glasgow to London.
6 Air services provide rapid transport from city centre to city centre.
7 Air services are not affected by weather conditions.
8 Short distance, bulk cargo is not suited to air transport.

B
1 Give two reasons why airports are situated away from the city centre.
2 Proposals to build a third London airport in addition to Heathrow and Gatwick are opposed by many people. Suggest three reasons for this.
3 Give two reasons why Post Office mail is an ideal cargo for aircraft.
4 Give three reasons why aircraft are not used to carry cargoes of coal.
5 Give two reasons why it would not be worthwhile to have a regular air service between Manchester and Hull.
6 Explain why aircraft take a larger proportion of passengers from London to Paris than from London to Glasgow (a greater distance).
7 Give two reasons why a passenger from London to Manchester may prefer to travel by rail rather than by air.

Route networks

The pattern of routes in an area is called a *transport network*. Fig. 125 shows three examples. Notice how their shapes vary according to the different methods of transport. Each network can be represented as a system of lines and points. The lines provide *links* between places; the points represent places or *nodes* (see Fig. 126).

Networks provide a set of routes along which the movement of goods and people takes place. Efficient networks allow a smooth and rapid flow of traffic from place to place. Many networks, however, cause long delays, low speeds and long detours. The road network of south Cheshire and north Shropshire, seen in Fig. 127, provides far superior communications for motorists than that of central Wales (Fig. 128).

An efficient transport network allows travellers to move easily to their destinations. Look at the road network which serves the imaginary island of Minim in Fig. 126. The island has five towns, all of which are linked together. Several journeys, however, can be made only by making long detours. The detour index of the journey from A to E, for example, is 285 – a journey nearly three times as long as the desire line distance (shortest path)

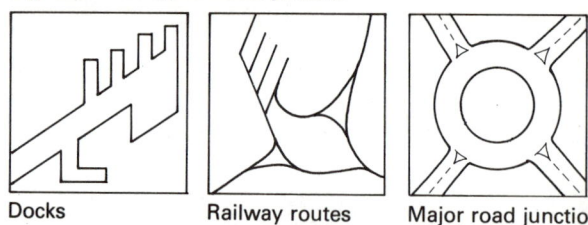

Fig. 125 Three typical route patterns

Docks   Railway routes   Major road junction

Fig. 126 The road network of 'Minim'

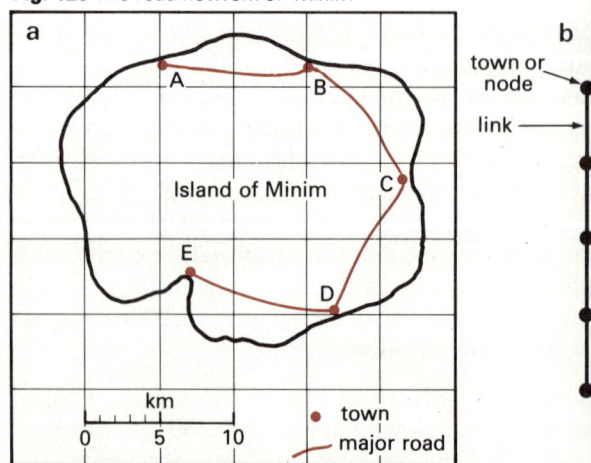

Island of Minim

km
0   5   10

● town
⌣ major road

a town or node
link

a The road network of the imaginary island of Minim
b A topological diagram of Minim's road network

from A to E. Traffic congestion is likely at C because journeys from A or B to D or E must all pass through C, even if there is no business at C.

A similar, but wealthier, island called Optim can afford a much better road network in which all five towns are directly linked to one another (Fig. 129). Each journey can be made without detour, and there are no traffic 'bottle-necks' to delay the movement of vehicles. Where roads cross, underpasses allow vehicles to maintain normal speed.

Topological diagrams

Useful comparisons of different transport networks can be made by simple mathematical techniques. To simplify calculations, route maps are transformed into *topological diagrams* (like the map of the London Underground). In a topological diagram, towns are shown as dots, and all routes are represented by straight lines. Figs. 126b and 129b show how the road maps for Minim and Optim appear as topological diagrams. (Remember that these diagrams do not give accurate scale or direction.)

Network density

The ease with which traffic can move from place to place often depends on the sheer number of roads in the area, that is the *density* of the road network. Fig. 127 clearly shows the greater density of the road network between Chester and Shrewsbury compared with that of central Wales (Fig. 128).

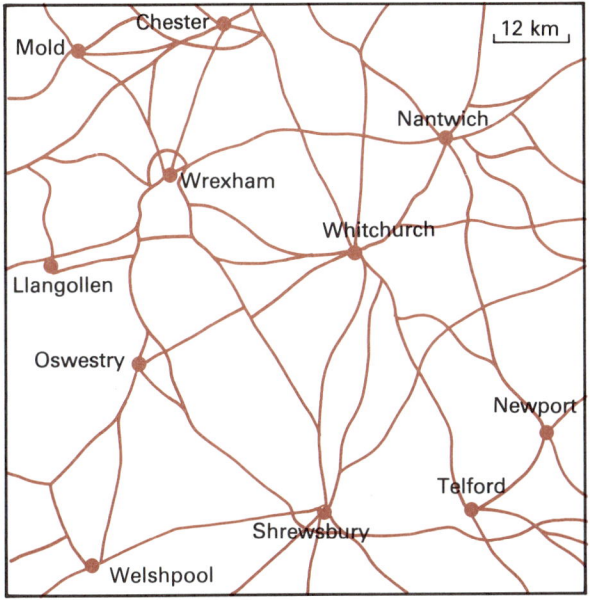

Fig. 127 The major road network of Cheshire and Shropshire

Fig. 128 The major road network of central Wales

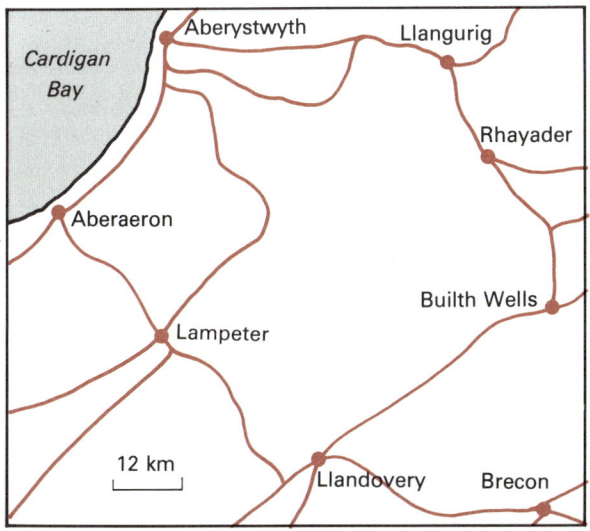

Fig. 129 The road network of 'Optim'

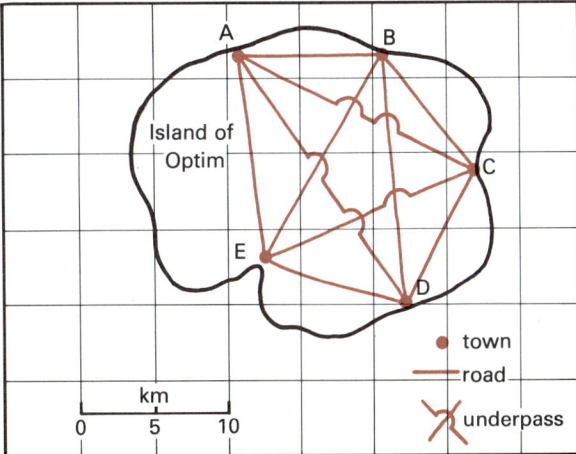

a Road network

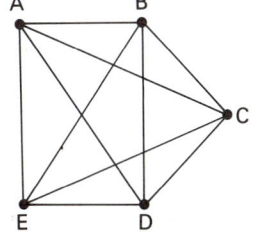

b Topological diagram

Areas of high relief and sparse population tend to have a low network density compared with a lowland area where road construction is cheaper and where there are more settlements to serve.

The *road density index* of a network is calculated according to the formula:

$$\frac{\text{total length of all roads in network}}{\text{total area covered by network}} \times \frac{100}{1}$$

The higher the index, the greater is the road density.

For Minim, the index is: $\dfrac{40 \text{ km}}{400 \text{ km}^2} \times \dfrac{100}{1} = 10$

Whereas that of Optim is: $\dfrac{140 \text{ km}}{400 \text{ km}^2} \times \dfrac{100}{1} = 35$

How well places are connected

How well places are connected to each other in the network as a whole is calculated by the *integration* (or *beta*) *index*. The towns of Optim, for example, are obviously better inter-connected than those of Minim. The formula is:

$$\text{integration index} = \frac{\text{number of links}}{\text{number of places}} \times \frac{100}{1}$$

Optim therefore has an integration index of $\dfrac{10}{5} \times \dfrac{100}{1} = 200$

Minim has an integration index of $\dfrac{4}{5} \times \dfrac{100}{1} = 80$

The higher the integration index, the better inter-connected are the towns in the network. Some towns may be better connected than others – that is, they are more *accessible* than others. To measure the accessibility of an individual town requires more detailed calculations.

Accessible places

An accessible place is well-connected by links to other nodes in the network. In other words, it is easy to reach from other places in the network. In Fig. 130, C is the most accessible place because it has direct links with each of the other four places. There are no other direct routes except those which focus on C.

Measuring accessibility

Fig. 131a shows an imaginary island with five towns. The government wishes to build a new fire station in the best position for crews to travel quickly to emergencies in any town. The most accessible town must be selected. This can be calculated by counting the number of links in the shortest route from place to place. In the top row of the matrix (Fig. 131b), for example, the journey from A to B takes just one link, whereas to travel from A to E involves three links. The total for each row shows how accessible each place is. The lowest row total, 5, indicates that C is the most accessible place. A is the least accessible town with a total of 9.

Fig. 130 An accessible place is well-connected by links

Fig. 131 Measuring accessibility

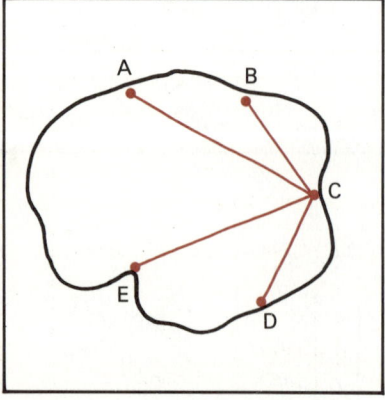

Which town is the most accessible?

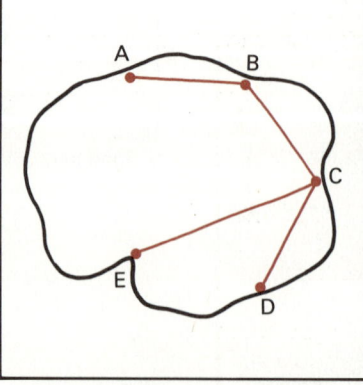

a Which is the most accessible place for a fire station?

	A	B	C	D	E	Total
A	—	1	2	3	3	9
B	1	—	1	2	2	6
C	2	1	—	1	1	5
D	3	2	1	—	2	8
E	3	2	1	2	—	8

b An accessibility matrix
A is least accessible
C is most accessible

The accessibility number

The *accessibility number* is a simple measure of how well a particular place is connected to other places. It is calculated by taking the shortest route to the most distant place in the network and counting the number of links needed to reach it. If we study the accessibility matrix in Fig. 131b, the furthest place from C is A, which can be reached in just two links. The accessibility number for C is therefore 2. The lower the accessibility number, the more accessible is the place concerned. In Fig. 130, C is the most accessible place with a number of 1; A, B, D and E each have an accessibility number of 2.

Now try Exercises 48, 49 and 50.

Exercise 48

Planning a road network

The five towns– Abbey, Boon, Cop, Dent and Eden– require a major new road network. The plans are of five different proposed systems. Each scheme was put forward by one of the following.

A: A travelling salesman who lives in Abbey and who makes daily visits to shops in the four towns – Boon, Cop, Dent and Eden – before returning home.

B: Abbey Council, which would like to make Abbey the busiest town in the area.

C: The Police, who need to be able to travel from one town to another as quickly as possible.

D & E: The County Council, for the area as a whole, which aims to spend as little money as possible. Proposal D caused a protest from two towns so plan E was put forward as an alternative to D.

1 (a) Match the proposed plans 1-5 to the proposers A, B, C, D and E.
   (b) Which two towns objected to plan D?

2 Calculate the total road distance for each of the five proposals.

3 The total area of the region is 3000 km².
   (a) Calculate the network density for each proposal.
   (b) Which is the most dense?
   (c) Which is the least dense?

4 (a) Calculate the integration index for plans 1, 2, 3 and 4.
   (b) Which of these networks is the most interconnected?
   (c) Which of the networks is the least interconnected?

5 Construct an accessibility matrix for plan 2 and one for plan 4. Which is the most accessible town in each case?

Proposed plans for road network

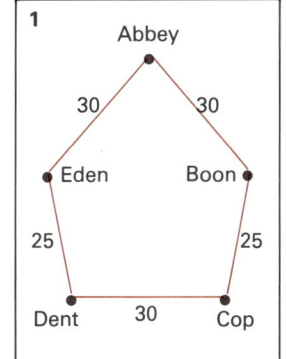

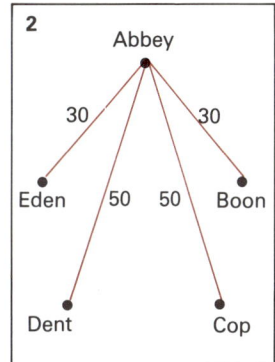

All roads are straight.

/ road

• town

15 / • journey distance between towns in kilometres

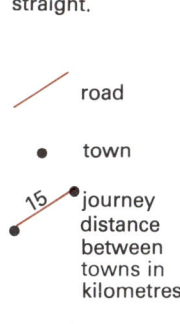

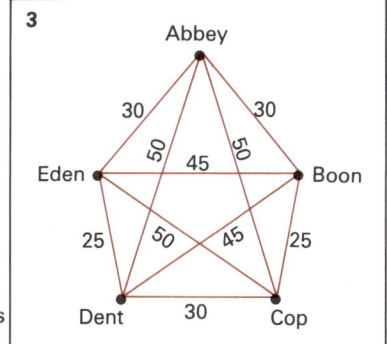

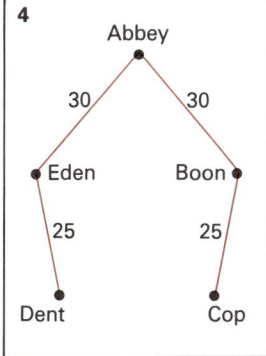

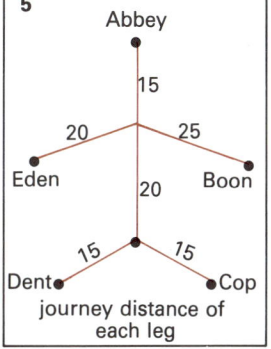

journey distance of each leg

Exercise 49 Playing away

Six teams play in the Anglican Football League. Teams play each other twice during the season, at home and away. Some teams travel much further than others. The journey matrix plots the distance each team travels each season.

1 Copy and complete the journey matrix.
2 Which team travels the greatest total distance?
3 Which team travels the shortest total distance?
4 Give (a) the shortest journey; (b) the longest journey.
5 Each season, all six teams enter a knock-out cup competition, the final of which is played at a stadium close to Wham's ground. Is this the most convenient location? Suggest which town might provide a more accessible location for the final.
6 If the roads on the map are the only roads, and there are no other towns,
   (a) Calculate Anglica's road network density if the total area is 60 000 km².
   (b) Calculate Anglica's road network integration index.

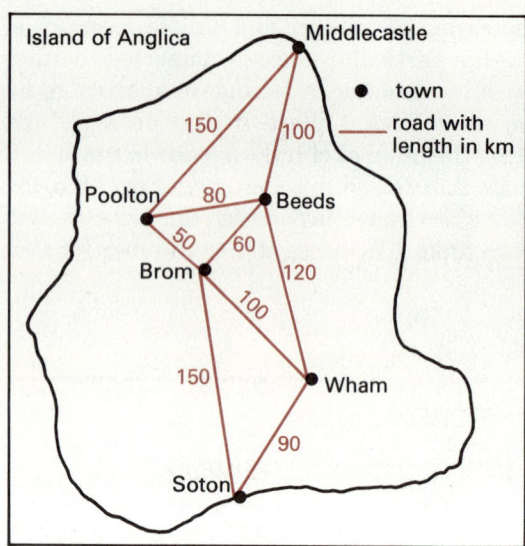

Island of Anglica

● town
— road with length in km

(c) Construct an accessibility matrix for towns in Anglica. Find from this: (i) the most accessible town; (ii) the least accessible town.

	Middlecastle	Beeds	Poolton	Brom	Wham	Soton	Total distance (km)
Middlecastle	×	100	150	160	220	310	940
Beeds		×					
Poolton			×				
Brom				×			
Wham					×		
Soton						×	

Exercise 50 Railway network

Study the map carefully and answer the questions.

1 Give two reasons why there are no railway services in area A.

2 Name the towns on the most direct railway route from London to Exeter.

3 Why have British Rail decided to develop the London–Exeter route via Bristol and not the most direct route?

4 Name the towns on the most direct main-line route from Glasgow to London which avoids Birmingham.

5 Name the towns on the most direct route from Newcastle upon Tyne to London.

6 Which of the following are termini:
(a) Exeter; (b) King's Lynn; (c) Bath; (d) Holyhead; (e) Stranraer?

7 Which junction serves the routes between:
(a) Newcastle, Preston and Glasgow;
(b) Manchester, Stafford, Chester and Stoke;
(c) London, Swindon, and Basingstoke;
(d) Newcastle, Leeds, Scarborough and Doncaster;
(e) Norwich, Cambridge and Lincoln?

8 A business conference is arranged for representatives from Hull, Preston, Exeter, Bristol, Liverpool, London, Leeds and Newcastle. Which of the following proposed centres is the most accessible by rail: London, Manchester, Derby or Birmingham?

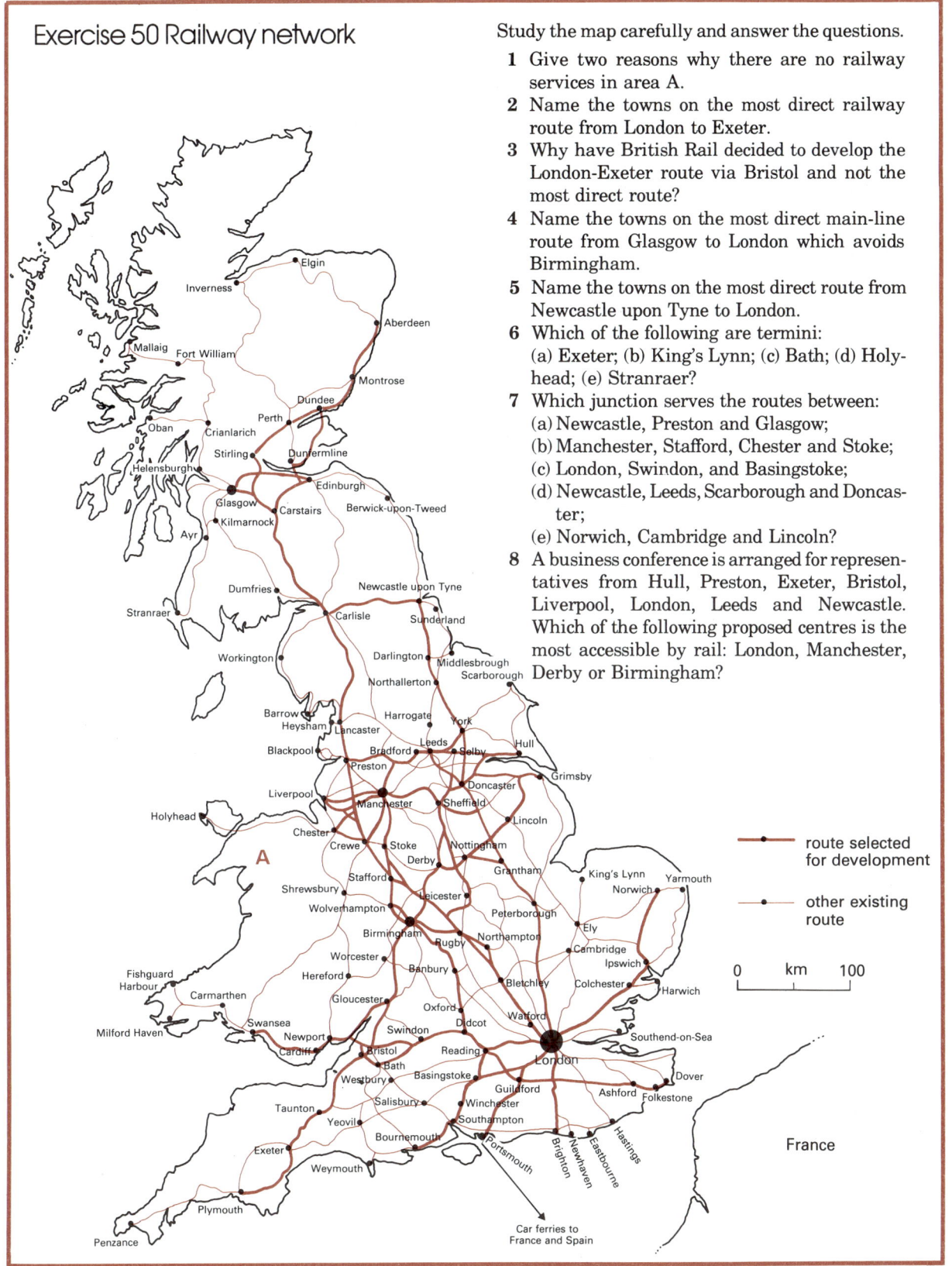

route selected for development

other existing route

0 km 100

France

Car ferries to France and Spain

Answers to Exercises

Exercise 1

1. **a** Fisherman, miner, market gardener, quarryman.
   b Baker, carpenter.
   c Bus driver, dentist, insurance clerk, shop assistant, disc jockey, professional footballer.
2. **a** Primary industry supplies raw materials.
   b Manufacturers make products from other goods.
   c Most British workers are employed in service industries.
3. B, C, D, E, J.
4. **a** farm wheat mill flour bakery bread shop
   b forest tree sawmill planks builder house
   c clay furnace bricks house
   d oil-well pipeline ship refinery road-tanker petrol station car
5. c.
6. **a** True, **b** True, **c** False, **d** True.

Exercise 2

1. Rich offshore fishing grounds; farming restricted by steep slopes and a short growing season.
2. Banned foreign fishing within 320 km of the Icelandic coast.
3. Tiny plants and animals which live in the sea.
4. Shallow sunlit waters; nearby river estuaries bringing nutrients.
5. Continental shelves.
6. Trawl-nets are draggged along the sea-bed, herring swim near the surface.
7. Demersal.
8. Trawl-nets; baited lines.
9. Newfoundland, Iceland.
10. Plaice — demersal; lobster — shellfish; crab — shellfish; mackerel — pelagic; pilchard — pelagic; herring — pelagic; halibut — demersal.

Exercise 3

1 True, 2 True, 3 False, 4 True, 5 False, 6 False, 7 False, 8 False, 9 False, 10 True.

Exercise 4

1. b, c, e.
2. To conserve fish stocks; to bring catch in line with demand.
3. The loss of deep-sea fishing grounds off Iceland, Greenland, Newfoundland and the Faroes.
4. Mackerel.
5. Control of mesh size for fishing nets; exclusion of foreign vessels from EEC waters; fixing of a maximum Total Allowable Catch.
6. Sixty percent of the Total Allowable Catch is taken from British territorial waters.

Exercise 5

1 True, 2 True, 3 False, 4 False, 5 False, 6 True, 7 False, 8 False, 9 False, 10 True.

Exercise 6

The baking of bread is a widespread *market-based* industry. Much weight is *gained* in the baking process because *water* is added to the mix. Bread is a bulky and *perishable* product which must be sold to *customers* while fresh. To keep *delivery* costs low, bakeries are found in most *towns*.

Exercise 7

1. bauxite alumina ingots rolling-mill cooking-foil
2. 1000 tonnes.
3. Hydroelectric power.
4. 170 kWh.
5. Cheap power supply from Kemano; the Douglas Channel provides a good harbour for shipping of raw materials and finished products.
6. **a** False, **b** False, **c** False, **d** False, **e** False.

Exercise 8

See table below.

Exercise 9

1. Sugar-beet and sugar cane.
2. 50%
3. 40 tonnes.
4. 3 tonnes.
5. harvesting slicing crushing boiling packing delivery
6. Water (steam) and pulp.
7. a, d.

Exercise 10

1. Because many other industries require steel, e.g. for making cars, aircraft and ships.
2. Coal.
3. **a** Strong to support the heavy overlying weight of iron-ore and limestone;
   b Porous to allow hot gases to pass through the blast-furnace.
4. iron-ore blast-furnace pig-iron oxygen converter rolling-mill steel girders
5. To create steels of additional strength or resistance to rust.
6. Electric arc furnace.
7. Blast-furnace.
8. An integrated steelworks combines blast-furnaces, steel-furnaces and rolling-mills on the same site.
9. Savings in fuel costs; savings in transport costs.
10. d.

Exercise 11

1 True, 2 True, 3 True, 4 True, 5 True, 6 True, 7 True, 8 True, 9 False (7800), 10 False (4500).

Exercise 12

1. Pottery — Stoke-on-Trent; glass — St. Helens; shipbuilding — Clydeside, Tyneside, Birkenhead, Belfast, Barrow. (Other examples: special steels — Sheffield; motor vehicles — Midlands; cotton textiles — Lancashire; woollen textiles — West Yorkshire.)
2. Local supplies of iron-ore; wood from nearby forests used to make charcoal for smelting; local sandstone used to make grindstones. (Also, fast-flowing Pennine streams provided water power.)
3. The tendency for an industry to remain in the area where it first developed even though the original reasons for its development are no longer important.
4. a, b, c.
5. Firms specialising in one particular process of an industry.
6. Loss of foreign customers because they are now making their own products; loss of customers because new industrial competitors such as Japan have increased their export business; the development of materials which can be used as substitutes

Exercise 8

Factory	Costs			
	Transport of beet	Processing beet	Transport of refined sugar	Total cost
A	NONE	£50	50p × 8 km = £4	£54
B	£4 × 5 km = £20	£50	50p × 5 km = £2.50	£72.50
C	£4 × 8 km = £32	£50	NONE	£82

Factory A is the most economical.

for steel, e.g., plastics, fibre-glass, aluminium, concrete.

7 a and b.

8 Chemicals, electrical goods, telecommunications, food processing.

9 South-east England.

10 Large market; labour supply.

Exercise 13

1 True, 2 False, 3 False, 4 False, 5 True, 6 True, 7 True, 8 False, 9 False, 10 True.

Exercise 14

A

1 Northern Ireland.
2 South East.
3 Northern Ireland.
4 South East.
5 South East.
6 East Anglia.
7 **a** North, North West, Wales, Scotland, Northern Ireland.
  b North, Yorkshire and Humberside, West Midlands, North West, Wales, Scotland, Northern Ireland.

B

1 Unemployed workers and empty factories are a waste of resources; the migration of workers seeking jobs in more prosperous regions causes housing shortages and overcrowding.

2 High costs of removal; higher costs of houses in areas of lower unemployment; social and family ties to the home area.

3 The infrastructure of an area is the quality of services such as housing, hospitals, roads, schools, power supplies, water and sewerage.

4 Because they have persistently above-average unemployment figures due to the decline of traditional industries such as coal-mining, steel manufacturing and shipbuilding.

5 To avoid these areas losing new developments to prosperous areas or to nearby Development Areas.

6 **a** South East, West Midlands, East Anglia.
  b Scotland.

Exercise 15

1 Coal and steel.
2 Over 75 000.
3 Because of a reduced demand for steel (reflecting foreign competition and world recession).
4 Car components, pharmaceuticals, electrical goods, confectionery, chemicals, oil refining, leather goods, plastics, printing. Service industries such as the Royal Mint and the Driver and Vehicle Licensing Centre.

5 Because the valleys are too narrow and inaccessible for large modern industrial estates.

6 Examples include the building of the M4 (including the Severn Bridge), the A465 'Heads of the Valleys Road', clearance and reclamation of derelict land.

7 Capital intensive industries, e.g. oil refining and chemicals, employ few workers.

8 Service industries offer a large number of jobs, especially those traditionally associated with female employment.

9 Modern growth industries may find suitable labour hard to find in South Wales; firms may be located more profitably in the South East closer to markets, and in an area with a more favourable infrastructure.

10 The new Royal Mint at Llantrisant; the Driver and Vehicle Licensing Centre at Swansea.

Exercise 16

1 **a** True, **b** False, **c** False, **d** True.
2 a, b, c, d, g.
3 Freedom from paying rates for ten years; total cost of factories, warehouses and offices to be paid by the State; Simplified planning regulations (other than Health and Safety).
  (Usual assistance available in Development Areas to be included.)
4 a, b, d.
5 a — 1; b — 1; c — 2; d — 2; e — 1; f — 0.

Exercise 17

1 Muscles provided the earliest form of power.
2 Muscle power depends upon food, which is a renewable source of energy.
3 Most electrical power is obtained by burning primary fuels such as coal, gas or oil.
4 Mechanical methods of production require increasing quantities of power; increase in world population requires a greater production of manufactured goods; motorised transport needs increasing amounts of oil-based fuel; increased living standards in the Developed World have increased the energy required for household appliances, air conditioning and central heating.
5 The world demand for energy is increasing; coal, oil and gas supplies are fixed and will eventually run out.
6 Ox, buffalo, donkey and horse.
7 Wind and water power.
8 Aircraft.
9 paddle sail steam-engine petrol engine diesel engine jet engine nuclear power
10 Coal was costly to transport.
11 Electricity.
12 a, b, c, d, e, f.
13 a, b, c, d, e, g, h, i.
14 a, b, c, e.
15 **a** Coal. **b** Oil. **c** More than doubled.

Exercise 18

The world faces an energy *shortage*. Reserves of *coal*, oil and gas are fixed and there is now a greater demand for power because world *population* and living standards are increasing. In poor countries such as India and *Brazil*, many people still use simple human and *animal* power to farm the land, for transport and to make *manufactured* goods. The energy needed for such power is obtained from *food*.

As living standards improve, factories use *electricity* to drive machines. Oil is needed for modern methods of transport. As *earnings* increase, people use more energy for *central-heating*, motor cars, and household goods such as televisions, washing-machines, vacuum cleaners and refrigerators.

Exercise 19

1 Coal, oil, natural gas.
2 Timber, solar power, tidal power, animal power, HEP.
3 Forest, peat, lignite, coal.
4 **a** Shaft mining. **b** Adit mining. **c** Opencast mining.
5 Flooding, fire, explosion, roof collapse.
6 Tanker, refinery.
7 Supplies of oil and natural gas are becoming scarcer. Coal reserves are sufficient for at least 300 years.
8 Coal is a non-renewable energy resource.

Exercise 20

1 Shaft mining at A and D; adit mining at B and C; opencast at G.
2 The shaft would have to be much deeper at D and would therefore be more costly to sink.
3 Coal seam is thicker at B; the valley side is less steep at B and extraction and transportation would be easier than at C.
4 The outcrop is much larger at G than at E and G is closer to the valleys and therefore nearer to possible lines of transport.
5 **a** True, **b** False, **c** True, **d** False, **e** False.

Exercise 21

1 crude oil — a thick, dark-coloured liquid extracted from an oil-well
2 drilling rig — a platform from which an oil-well is drilled
3 oil refinery — a factory which splits crude oil into separate hydrocarbons
4 paraffin — a hydrocarbon produced at an oil refinery
5 plankton — microscopic plants and animals which live in the sea

Answers to Exercises

6 fault —   a crack in the Earth's crust

7 cap rock —   a layer of rock which covers a reservoir of oil and gas

8 hydrocarbons —chemical substances which make up crude oil

Exercise 22

1 False, 2 False, 3 True, 4 False, 5 False, 6 False, 7 False, 8 False, 9 False, 10 False, 11 False, 12 True.

Exercise 23

1 See diagram opposite.
2 Coal (also lignite and peat), natural gas, oil.
3 Hydroelectric power-station.
4 e.
5 High-pressure steam.
6 heat   water   steam   turbine generator   electricity   transmission consumer
7 National Supergrid.
8 Primary fuels such as coal, oil and natural gas are burnt to produce the steam needed to generate electricity, which is therefore a secondary form of power.
9 Heat is lost during the process of generating electricity; electricity cannot be stored efficiently, and as many power-stations are required to supply peak demand only, they remain idle for much of the day.
10 Because they are required for peak demand only.
11 d, e.
12 Hydroelectric power; wind power; wave power.

Exercise 24

1 C is too distant from the coalfield and the towns; transporting coal to the power-station and erecting transmission lines to the towns would be too costly because it is separated by a mass of highland.
2 The advantages of site A are: its close proximity to the towns and the coalfield; easy access to the railway for transporting coal. The disadvantages are: the prevailing winds would blow the fumes into the town; site A is away from the river and so from important cooling water.
3 a and b.
4 B is situated on the coalfield; cooling water is available from the river; the site is centrally placed for three of the four towns; the site is in a lowland area with sufficient flat land on which to construct the power-station.

Exercise 25

The *water-wheel* was used to produce power to drive *flour-mills* and *machines* during the early part of the Industrial Revolution. Today *hydroelectric* power-stations use the *force* of moving water to drive *turbines*. HEP stations are best situated in mountainous areas of regular and plentiful *rainfall*. Although *operating* costs are low, installation costs are expensive. Because it is a *renewable* energy resource, HEP is likely to become more important as supplies of oil become *scarce*.

Exercise 26

1 Site A is close to the major towns, transmission costs would be low; it is on a large river fed by several tributaries. Building a dam to create a sufficient head of water would be difficult in such a lowland area.
2 a C; b D; c C; d A.
3 a Site B is in a steep-sided valley suitable for building a dam to produce a good head of water; it is supplied by several tributaries; it is easily accessible and not too distant from the towns on the eastern coast; transmission costs would be low.
b The high cost of transmitting electricity to the town on the northern coast; high construction costs of dam.

Exercise 27

1 True, 2 False, 3 False, 4 True, 5 True, 6 True, 7 True, 8 False, 9 False, 10 False, 11 False, 12 False.

Exercise 28

1 a thermal —   burns coal or oil
   b solar —   uses mirrors
   c HEP —   depends on rainfall
   d nuclear —   needs a reactor
   e tidal —   is always sited on the coast
   f geothermal —uses underground heat
2 a Uranium (others are oil-based).
   b West Burton (others are nuclear).
   c HEP (others are non-renewable).
   d Oil (others are plant-based).
3 a petrol, b paraffin, c uranium, d water, e coal, f sun, g sea, h wind.
4 b and c.
5 Nuclear.
6 Coal, peat, lignite, oil, natural gas, uranium.
7 Coal, peat, lignite, oil, natural gas.
8 HEP, wind, tidal, wave, solar, geothermal, manpower, animals.
9 Organisation of Petroleum Exporting Countries.
OPEC controls 60% of the world's oil supplies; many of the world's industrial

nations are dependent upon OPEC for supplies of oil.
10 a Roof insulation; cavity-wall insulation; draught-proofing windows and doors; double glazing.
b Cars designed to reduce petrol consumption; increase number of passengers per bus or train so that journeys made by private car are reduced; improve facilities for pedestrians and cyclists.
11 b and d.
12 Coal, oil and natural gas.

Exercise 29

1 See table opposite.
2 a Bristol. b Newport, Chepstow, Gloucester.
c Chepstow, Bristol. d Gloucester.
e Bristol, Chepstow, Newport.
3 Any two of the journeys to Swindon from Hereford, Chepstow, Newport, Cardiff, Merthyr Tydfil.
4 a 80 km. b 1 hour in time; 8 litres of petrol.
5 a $\frac{226}{53} \times \frac{100}{1} = 426.41$.
b $\frac{146}{53} \times \frac{100}{1} = 275.47$.

Exercise 30

1 Route b (735 m).
2 A → C → building society → E → department store → E → dress shop → C → A (635 m).

OR

A → C → dress shop → E → department store → E → building society → C → A (635 m).

Exercise 31

1 See table opposite.
2 a True, b True, c True, d False.
3 a 16%, b 36%, c 28%.
4 All of them will affect the type of transport chosen.

Exercise 32

A   1 Air (air mail).
     2 Ocean tanker.
     3 Pipeline.
     4 Underground railway ('tube').
     5 Railway.
     6 Road tanker.
     7 Railway.
     8 On foot.
     9 On foot/bicycle.
     10 Electric van/milk float.
B   1, 4, 7, 8, 9, 10.

Answers to Exercises

Exercise 23 1

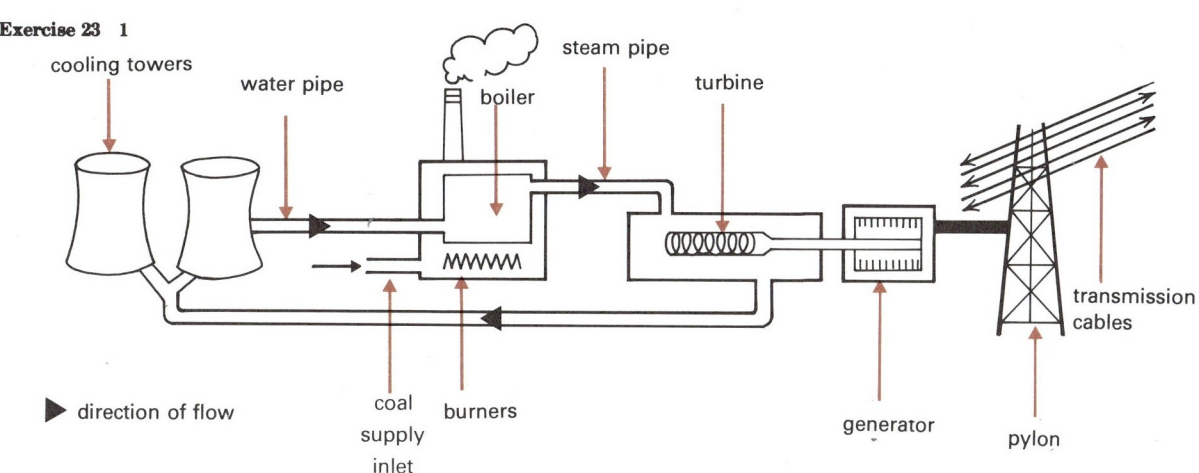

cooling towers · water pipe · steam pipe · boiler · turbine · transmission cables · coal supply inlet · burners · generator · pylon

▶ direction of flow

Exercise 29 1	Bristol		Bath		Swindon		Taunton	
	Via Glos	Via Brge	Via Glos	Via Brge	Via Glos	Via Brge	Via Glos	Via Brge
Hereford	106	79	127	100	105	166	183	156
Chepstow	101	21	122	42	100	108	178	98
Newport	128	48	149	69	127	135	205	125
Cardiff	149	69	170	90	148	156	226	146
Merthyr	188	108	209	129	187	195	265	185

Exercise 31 1

MODE	Up to 1 km	1–2 km	2–3 km	Over 3 km	TOTAL
Walk	7	0	0	0	7
Bicycle	1	3	3	2	9
Bus	0	1	1	2	4
Car	0	0	1	4	5
TOTAL	8	4	5	8	25

Distance of pupil's home from school

Exercise 33 1

Mode	Average speed km/h	Time hours	
		Travelling	London
bus	50	8	1
car	80	5	4
train	100	4	5

Exercise 33

1 See table below.
2 **a** 11.30; **b** 11.00.
3 **a** 15.30; **b** 16.00.
4 b, c, d, e, g, h, i, j.
5 **a** 8 hours. **b** 09.30.

Exercise 34

1 c.
2 d.
3 To reduce financial loss made by British Railways.
4 Closure of unprofitable lines; modernisation of key intercity routes; introduction of diesel and electric locomotives; introduction of freightliner goods services.
5 d.
6 b and c.

Exercise 35

1 False, **2** True, **3** True, **4** True, **5** False, **6** True, **7** False, **8** True, **9** False, **10** False.

Exercise 36

1 They provide the following services: parking facilities, fuel, repairs, food and drink, toilets, rest.
2 Motor vehicles cannot cross the carriageway.
3 Personal services for the motorist and passengers, e.g. restaurant, rest areas.
4 d.
5 A — slip road from motorway; B — service road to centre; C — central barrier; D — bridge and restaurant; E — access road to motorway.
6 X is the parking area for cars, Y is for transport vehicles, e.g., lorries, vans, coaches.
7 Noise; air pollution; problems of access to separated parts of the farm; loss of farm land needed for motorway construction.
8 d, e, f.

Answers to Exercises

Route	Distance *km*			Time		
B to A	'A' class road	Motorway	Total	'A' class road	Motorway	Total
'A' class road only	100	—	100	2h	—	2h
Via motorway and 'A' class road	110	130	240	2h 12 min	1h 18 min	3½h
D to B						
'A' class road only via A	205	—	205	4h 6 min	—	4h 6 min
Via motorway and 'A' class road	75	130	205	1h 30 min	1h 18 min	2h 48 min

Exercise 37

1 See table above.
2 a None.
  b D to B; D to C; A to C.

Exercise 38

1 See table right.
2 A 190km. B 220km. C 230km.
3 Detour will reduce total cost of route A by £11m to £78m.
4 a £42m. b £52m. c £48m.
5 B $\frac{220}{190} \times \frac{100}{1} = 115.78.$

  C $\frac{230}{190} \times \frac{100}{1} = 121.05.$

  X 100.

  Y $\frac{73}{53} \times \frac{100}{1} = 137.74.$
6 X is shorter than Y thereby reducing journey time;
  X will cross the highland which will be less populated and have fewer alternative uses;
  Y would cross the lowland area which is more likely to be used for farming and housing.

Exercise 39

1 a A. b B. c D. d E. e C.
2 a A, C, E. b C, E. c C, E. d C.
3 C, D, E.
4 A, B, D, E.

Exercise 40

1 The Romans. To enable soldiers to move quickly between forts and towns.
2 A straight road is the most direct (shortest route).
  A straight road might encounter steep slopes, marshland or stretches of open

Exercise 38 1

Route	Cost of land	Construction costs	Total costs
A	£34m	£55m	£89m
B	£44m	£37m	£81m
C	£44m	£48m	£92m

water, or might require the demolition of buildings.
3 Most traffic went by rail during this period.
4 The increase in numbers of road vehicles may be greater than road improvements; better long-distance roads increase congestion as vehicles approach built-up areas with traditional roads.
5 Pedestrians, horses and bicycles are banned; limited points of access from slip roads; two or three lanes allow traffic to flow at different speeds; gradients reduced by cuttings and embankments; gentle bends allow vehicles to maintain high speed.
6 A motorway carrying traffic through a built-up area.
7 High cost of urban land needed for motorway construction; noise and air pollution faced by local residents; physical disruption in and division of local neighbourhoods.
8 Insufficient year-round traffic to justify construction; relief and indented coastline would make construction very costly.
9 Fog.
10 a by-pass — a detour for through traffic
   b pedestrian precinct — a walking area for shoppers
   c clearway — stops parking on main roads
   d multi-storey car-park — increases parking spaces in city centres
   e lay by — permits roadside parking without holding up traffic
   f bus lane — a method of increasing speed of public transport
   g underpass — separates traffic into two levels
   h tidal traffic flow — reverses direction of vehicles in central road lanes
   i roundabout — causes circular flow of traffic
   j no right turns — improves traffic flow along central lanes at road junctions

Exercise 41

1 Into London.
2 A — morning rush-hour; B — lunch-hour; C — workers returning home; D — visitors to clubs and cinemas.
3 a morning rush-hour between 08.00 and 09.00; b 04.00.
4 a, d.
5 a, c, d.

Exercise 42

1 a 40; b 10; c 60; d 100.
2 290 travelled towards junction, therefore 290 travelled away.
3 a North Street, with a two-way traffic volume of 205 vehicles.
  b East Street, with a two-way traffic volume of 95 vehicles.
4 a Facing South Street; b facing East Street.
5 North.

Exercise 43

1 Completed traffic flow diagram showing the following routes:

Tom → A → D → E → Town Hall.
Jack → A → D → E → railway station.
Jane → B → C → D → E → Town Hall.
Bill → C → D → E → Marks and Spencer.
Ian → G → F → D → E → railway station.
Roy → G → F → D → E → Marks and Spencer.
Eve → H → F → D → E → railway station.
Pam → H → F → D → E → Marks and Spencer.
Paul → H → F → D → E → Town Hall.

2 A — 2; B — 2; D — 10; E — 10.

3 D—E, F—D, H—F, G—F.

4 a 2; b 4; c 2.

5 Ian, 7½km.
Bill, 4½km.

6 a $\frac{4.5}{4} \times \frac{100}{1} = 112.5$.

b $\frac{7}{2} \times \frac{100}{1} = 350$

7 a Paul, Pam, Eve, Roy, Ian.
b None.
c i Paul's journey would be 3½km, half his original journey.
ii Paul's detour index would be halved to 175.

Exercise 44

1 dug-out canoe   sailing-ship   horse-drawn barge   paddle steamer   turbine-driven liner   nuclear submarine   hovercraft

2 It is low cost; heavy, bulky cargoes may be carried.

3 Low speed; winter freezing; dredging may be needed to maintain navigable channels.

4 The River Rhine.

5 From the mouth of the Rhine (Rotterdam) upstream to Basel.

6 A coalfield region with heavy industry in northern West Germany.

7 The McKenzie flows northwards to the empty Arctic region; winter freezing and spring flooding restricts navigation.

8 a Summer drought reduces water depth.
b Winding course (meanders) increases voyage length.
c Winter freezing, spring floods.

9 Panama, Suez, St Lawrence Seaway, Manchester Ship Canal.

10 Level but winding course; locks; narrow width reduces speed; winter freezing.

11 They have been replaced by railway and road transport.

12 Clay, coal and sand.

13 Barges can complete a journey from one inland location, across the sea locked to a catamaran, and thence to an inland destination abroad. Cargoes therefore remain undisturbed.

14 Manchester Ship Canal.

15 Chemicals, tar, soap, margarine and other food processing.

Exercise 45

Ocean tankers provide the *cheapest* means of transporting *oil*. Vessels have been increased in size to reduce transport *costs* per tonne. Special deep-water *terminals* are needed to accommodate *supertankers* of over 300 000 *tonnes*. Because such huge vessels cannot *dock* at *Fawley* or even Milford Haven, they must *unload* their cargoes at *Bantry Bay*, from where smaller tankers can deliver oil to British ports.

Exercise 46

a Liners — passenger ships severely hit by air transport since 1950.
b Cross-channel ferries — offer a shuttle service to the continent.
c Tramp cargo ships — their route, time-table and cargo varies according to the business available.
d Bulk-carriers — large vessels built for carrying a specific cargo such as wheat or oil.
e Freightliners — ships carrying assorted cargo over regular long distance routes.
f Containers — can be carried by road, rail and sea transport.
g Roll-on/roll-off ships — carry lorries loaded with goods across the Channel.
h BACAT barge — an ocean-going vessel which carries canal barges.

Exercise 47

A 1 True, 2 True, 3 False, 4 False, 5 False, 6 False, 7 False, 8 True.

B 1 Large areas of land are needed, and are not available near city centres; it is safer to build airports away from built-up areas; noise nuisance is reduced.

2 Local residents living near the proposed site put forward the following objections: loss of valuable farm land; nuisance from noise; increase in traffic and commercial activities disrupting local village communities.

3 Light weight, low bulk; high value; urgent delivery required.

4 High weight and bulk, low value.

5 Manchester and Hull are well-served by motorway (M62) and by rail services (inter-city); there is insufficient passenger traffic to justify regular air services over such a short distance (140km).

6 London to Glasgow is a rapid inter-city railway route (5-hour journey); London to Paris by a journey other than air requires a slow channel crossing.

7 Rail journey is cheaper; it may be quicker from city centre to city centre; aircraft are more likely to be delayed by bad weather; some passengers dislike air travel.

Exercise 48

1 a 1 — A, 2 — B, 3 — C, 4 — D, 5 — E.
b Dent and Cop.

2 1 — 140km, 2 — 160km, 3 — 380km, 4 — 110km, 5 — 110km.

3 a 1 — 4.66, 2 — 5.33, 3 — 12.66, 4 — 3.66, 5 — 3.66.
b 3.
c 4 and 5.

4 a 1 — 100, 2 — 80, 3 — 200, 4 — 80.
b 3.
c 2 and 4.

5 See tables below.

Exercise 48 5

PLAN 2

	Abbey	Boon	Cop	Dent	Eden	TOTAL
Abbey	—	1	1	1	1	4
Boon	1	—	2	2	2	7
Cop	1	2	—	2	2	7
Dent	1	2	2	—	2	7
Eden	1	2	2	2	—	7

PLAN 4	Abbey	Boon	Cop	Dent	Eden	TOTAL
Abbey	—	1	2	2	1	6
Boon	1	—	1	3	2	7
Cop	2	1	—	4	3	10
Dent	2	3	4	—	1	10
Eden	1	2	3	1	—	7

Abbey is the most accessible town in each case.

Answers to Exercises

	Middlecastle	Beeds	Poolton	Brom	Wham	Soton	Total distance (km)
Middlecastle	x	100	150	160	220	310	940
Beeds	100	x	80	60	120	210	570
Poolton	150	80	x	50	150	200	630
Brom	160	60	50	x	100	150	520
Wham	220	120	150	100	x	90	680
Soton	310	210	200	150	90	x	960

Exercise 49

1 See table above.
2 Soton.
3 Brom.
4 **a** Poolton to Brom. **b** Middlecastle to Soton.
5 No. Brom.

6 **a** $\dfrac{900}{60\ 000} \times \dfrac{100}{1} = 1.5.$
   b $\dfrac{9}{6} \times \dfrac{100}{1} = 150.$
   c See table below.

	Middlecastle	Beeds	Poolton	Brom	Wham	Soton	TOTAL
Middlecastle	—	1	1	2	2	3	9
Beeds	1	—	1	1	1	2	6
Poolton	1	1	—	1	2	2	7
Brom	2	1	1	—	1	1	6
Wham	2	1	2	1	—	1	7
Soton	3	2	2	1	1	—	9

i Brom and Beeds; **ii** Middlecastle and Soton.

Exercise 50

1 Area A is Snowdonia which is a mountainous region where
   a railways are difficult to build and
   b there are few people and no large towns.
2 Basingstoke, Salisbury, Yeovil.
3 The London to Exeter via Bristol route joins the east to the west linking the major towns of Reading, Swindon and Bristol and also links routes from South Wales with London.
4 Carstairs, Carlisle, Lancaster, Preston, Crewe, Stafford, Rugby, Bletchley, Watford.
5 Darlington, Northallerton, York, Doncaster, Grantham, Peterborough.
6 b, d, e.
7 **a** Carlisle; **b** Crewe; **c** Reading; **d** York; **e** Ely.
8 Birmingham.

Index

Index